Advanced Practical Organic Chemistry

Dedicated to Professor Gilbert Stork
In recognition of the skills and enthusiasm for chemistry
gained in his laboratories

Advanced Practical Organic Chemistry

Second edition

J. LEONARD
B. LYGO
G. PROCTER
Department of Chemistry and Applied Chemistry
University of Salford

BLACKIE ACADEMIC & PROFESSIONAL
An Imprint of Chapman & Hall

London · Glasgow · Weinheim · New York · Tokyo · Melbourne · Madras

**Published by Blackie Academic & Professional, an imprint of
Chapman & Hall, Wester Cleddens Road, Bishopbriggs, Glasgow G64 2NZ**

Chapman & Hall, 2-6 Boundary Row, London SE1 8HN, UK

Blackie Academic & Professional, Wester Cleddens Road, Bishopbriggs,
Glasgow G64 2NZ, UK

Chapman & Hall GmbH, Pappelallee 3, 69469 Weinheim, Germany

Chapman & Hall USA, One Penn Plaza, 41st Floor, New York NY 10119,
USA

Chapman & Hall Japan, ITP-Japan, Kyowa Building, 3F, 2-2-1 Hirakawacho,
Chiyoda-ku, Tokyo 102, Japan

DA Book (Aust.) Pty Ltd, 648 Whitehorse Road, Mitcham 3132, Victoria,
Australia

Chapman & Hall India, R. Seshadri, 32 Second Main Road, CIT East, Madras
600 035, India

First edition 1990
Second edition 1995

© 1995 Chapman & Hall

Printed in Great Britain by The Alden Press, Osney Mead, Oxford

ISBN   0 7514 0200 1

A catalogue record for this book is available from the British Library

Library of Congress Catalog Card Number: 94-72928

∞ Printed on acid-free text paper, manufactured in accordance with
   ANSI/NISO Z39.48-1992 (Permanence of Paper).

Preface to Second Edition

The preparation of organic compounds is central to many areas of scientific research, from the most applied to the most academic, and is not limited to chemists. Any research which uses new organic chemicals, or those which are not available commercially, will at some time require the synthesis of such compounds.

This highly practical book continues to cover the most up-to-date techniques commonly used in organic synthesis and is based on our experience of establishing research groups in synthetic organic chemistry and our association with some of the leading laboratories in the field. It is not claimed to be a comprehensive compilation of information to meet all possible needs and circumstances; rather, the intention has been to provide sufficient guidance to allow the researchers to carry out reactions under conditions which offer the highest chance of success.

The book is written for postgraduate and advanced level undergraduate organic chemists and for chemists in industry, particularly those involved in pharmaceutical, agrochemical and other fine chemicals research. Biologists, biochemists, genetic engineers, material scientists and polymer researchers in university and industry will find the book a useful source of reference.

All the artwork for this second edition has been modernized and the chapters have been reorganized to allow for more expanded treatment of practical techniques.

Contents

1 General Introduction   **1**

2 Safety   **3**

2.1   Safety is your primary responsibility   3
2.2   Safe working practice   4
2.3   Common hazards   4
2.4   Accident and emergency procedures   6
2.5   Bibliography   7

3 Keeping records of laboratory work   **8**

3.1   Introduction   8
3.2   The laboratory notebook   8
    3.2.1   Why keep a lab book?   8
    3.2.2   How to write a lab book   9
    3.2.3   Suggested notebook format   10
3.3   Keeping records of data   13
    3.3.1   Purity, structure determination and characterization   13
    3.3.2   What type of data should be collected   14
    3.3.3   Formats for data records   19
3.4   Some tips on report and thesis writing   24
    3.4.1   Sections of a report or thesis   24
    3.4.2   Planning a report or thesis   24
    3.4.3   Writing the report or thesis   26

4 Equipping the laboratory and the bench   **36**

4.1   Introduction   36
4.2   Setting up the laboratory   36
4.3   General laboratory equipment   37

	4.4	The individual bench	44
	4.4.1	Routine glassware	44
	4.4.2	Personal items	45
	4.4.3	Specialized personal items	46

5 Purification and drying of solvents **54**

	5.1	Introduction	54
	5.2	Purification of solvents	54
	5.3	Drying agents	55
	5.4	Drying of solvents	59
	5.5	Solvent stills	65

6 Reagents: Preparation, purification and handling **69**

	6.1	Introduction	69
	6.2	Classification of reagents for handling	70
	6.3	Techniques for obtaining pure and dry reagents	71
	6.3.1	Purification and drying of liquids	71
	6.3.2	Purifying and drying solid reagents	74
	6.4	Techniques for handling and measuring reagents	76
	6.4.1	Storing liquid reagents or solvents under inert atmosphere	76
	6.4.2	Bulk transfer of a liquid under inert atmosphere (cannulation)	78
	6.4.3	Using cannulation techniques to transfer measured volumes of liquid under inert atmosphere	81
	6.4.4	Use of syringes for the transfer of reagents or solvents	84
	6.4.5	Handling and weighing solids under inert atmosphere	92
	6.5	Preparation and titration of simple organometallic reagents	98
	6.5.1	General considerations	98
	6.5.2	Preparation of Grignard reagents	99
	6.5.3	Preparation of organolithium reagents	100
	6.5.4	Titration of lithium reagents	102
	6.5.5	Preparation of lithium amide bases	102
	6.6	Preparation of diazomethane	103
	6.6.1	Safety measures	104
	6.6.2	Preparation of diazomethane (a dilute ethereal solution)	104
	6.6.3	General procedure for esterification of carboxylic acids	106

    6.6.4   Titration of diazomethane solutions   106

7   Gases   107

7.1   Introduction   107
7.2   Use of gas cylinders   107
       7.2.1   The procedure for fitting a regulator to a cylinder   109
7.3   Handling gases   110
7.4   Measurement of gases   113
       7.4.1   Measurement of a gas using standardized solution   113
       7.4.2   Measurement of a gas using a gas tight syringe   113
       7.4.3   Measurement of a gas using a gas burette   114
       7.4.4   Quantitative analysis of hydride solutions using a gas burette115
       7.4.5   Measurement of a gas by condensation   116
       7.4.6   Measurement of as gas using a quantitative reaction   117
7.5   Inert gases   117
7.6   Reagent gases   118
       7.6.1   Gas scrubbers   119
       7.6.2   Methods for preparing some commonly used gasses   119

8   Vacuum pumps   122

8.1   Introduction   122
8.2   House vacuum systems   122
8.3   Medium vacuum systems   123
       8.3.1   Water aspirators   123
       8.3.2   Electric diaphragm pumps   123
8.4   High vacuum pumps   124
       8.4.1   Rotary oil pumps   124
       8.4.2   Vapour diffusion pumps   126
8.5   Pressure measurement and regulation   126
       8.5.1   Units of pressure (vacuum) measurement   127

9   Carrying out the reaction   128

9.1   Introduction   128
9.2   Reactions with air sensitive reagents   129

9.2.1   Introduction   129

9.2.2   Preparing to carry out a reaction under inert conditions   129

9.2.3   Drying and assembling glassware   130

9.2.4   Typical reaction set-ups using a double manifold   131

9.2.5   Basic procedure for inert atmosphere reactions   132

9.2.6   Modifications to basic procedure   135

9.2.7   Use of balloons for holding an inert atmosphere   140

9.2.8   The use of a 'spaghetti' tubing manifold   143

9.3   Reaction monitoring   144

9.3.1   Thin layer chromatography   145

9.3.2   High performance liquid chromatography (hplc)   152

9.3.3   Gas-liquid chromatography (gc, glc, vpc)   156

9.4   Reactions at other than room temperature   159

9.4.1   Low temperature reactions   160

9.4.2   Reactions above room temperature   163

9.5   Driving equilibria   168

9.5.1   Dean and Stark traps   169

9.5.2   High pressure reactions   170

9.6   Agitation   170

9.6.1. Magnetic stirring   170

9.6.2. Mechanical stirrers   172

9.6.3. Mechanical shakers   174

9.6.4. Sonication   175

10 Working up the reaction   177

10.1   Introduction   177

10.2   Quenching the reaction   178

10.2.1 General comments   178

10.2.2 Strongly basic non-aqueous reactions   178

10.2.3 Neutral non-aqueous reactions   179

10.2.4 Strongly acidic non-aqueous reactions   180

10.2.5 Acidic or basic aqueous reactions   180

10.2.6 Liquid ammonia reactions   180

10.3   Isolation of the crude product   181

10.3.1 General comments   181

10.3.2 Very polar aprotic solvents   182

11 Purification   184

11.1	Introduction	184
11.2	Crystallization	184
	11.2.1 Simple crystallization	184
	11.2.2 Small scale crystallization	187
	11.2.3 Crystallization at low temperature	189
	11.2.4 Crystallization of air sensitive compounds	192
11.3	Distillation	193
	11.3.1 Simple distillation	193
	11.3.2 Distillation under inert atmosphere	194
	11.3.3 Fractional distillation	196
	11.3.4 Distillation under reduced pressure	197
	11.3.4 Small scale distillation	201
11.4.	Sublimation	203
11.5	Chromatographic purification of reaction product	204
11.6	Flash chromatography	205
	11.6.1 Equipment required for flash chromatography	205
	11.6.2 Procedure for running a flash column	208
	11.6.3 Recycling procedure for flash chromatography	214
11.7	Dry-column flash chromatography	215
	11.7.1 Method for running a dry flash column	215
11.8	Preparative tlc	217
11.9	Medium pressure liquid chromatography	217
	11.9.1 Setting up an mplc system	218
	11.9.2 Procedure for using mplc	221
11.10	Preparative hplc	224
11.10.1	Equipment required	224
11.10.2	Running a preparative hplc column	224

12 Small scale reactions   227

12.1	Introduction	227
12.2	Reactions at or below room temperature	228
12.3	Reactions above room temperature	231
12.4	Reactions in nmr tubes	232
12.5	Purification of materials	233

	12.5.1	Distillation	233
	12.5.2	Crystallization	234
	12.5.3	Chromatography	234

13 Large scale reactions — **236**

13.1	Introduction	236
13.2	Carrying out the reaction	237
13.3	Purification of the products	237

14 Special procedures — **240**

14.1	Introduction	240
14.2	Catalytic hydrogenation	240
14.3	Photolysis	244
14.4	Ozonolysis	246
14.5	Flash vacuum pyrolysis	247
14.6	Liquid ammonia reactions	247

15 Characterization — **250**

15.1	Introduction	250
15.2	Nmr	251
15.3	Ir	253
15.4	Uv	254
15.5	Mass spectra	254
15.6	M.p. and b.p.	254
15.7	Optical rotation	255
15.8	Microanalysis	256
15.9	Keeping the data	257

**16 'Trouble shooting';
what to do when things don't work** — **258**

17 The chemical literature — **262**

| 17.1 | The structure of the chemical information | 262 |

	17.1.1	Introduction	262
	17.1.2	The structure of the literature	263
17.2		Some important paper-based sources of chemical information	264
	17.2.1	Chemical Abstracts	264
	17.2.2	Beilstein	265
	17.2.3	Science Citation Index	266
17.3		Some important electronic-based sources of chemical information	267
	17.3.1	Beilstein	269
	17.3.2	CAS ON-LINE	269
	17.3.3	CASREACT	269
	17.3.4	SCI	269
	17.3.5	Chemical Journals On-line (CJO)	270
	17.3.6	REACCS (Molecular Design Ltd.)	270
	17.3.7	Cambridge Structure Database	270
	17.3.8	The World Wide Web	270
17.4		How to find chemical information	271
	17.4.1	How to do searches	271
	17.4.2	How to find information on specific compounds	272
	17.4.3	How to find information on classes of compounds	273
	17.4.4	How to find information on synthetic methods	274
17.5		Current awareness	275

Appendices **277**

Appendix 1.	Properties of common solvents	277
Appendix 2.	Properties of common gases	278
Appendix 3.	Approximate pKa values for some common deprotonations compared to some common bases	279
Appendix 4.	Lewis acids	280
Appendix 5.	Common reducing reagents	281
Appendix 6.	Common oxidizing reagents	285

Index **289**

CHAPTER 1

General Introduction

The preparation of organic compounds is central to many areas of scientific research, from the most applied to the most "academic", and is not limited to chemists. Any research which uses new organic chemicals, or those which are not available commercially, will at some time require the synthesis of such compounds. Accordingly the biologist, biochemist, genetic engineer, materials scientist, and polymer researcher in university or industry all might find themselves faced with the task of carrying out an organic preparation, along with those involved in pharmaceutical, agrochemical, and other fine chemicals research.

These scientists share with the new organic chemistry graduate student a need to be able to carry out modern organic synthesis with confidence and in such a way as to maximize the chance of success. The techniques, methods, and reagents used in organic synthesis are numerous, and increasing every year. Many of these demand particular conditions and care at several stages of the process, and it is unrealistic to expect an undergraduate course to prepare the chemist for all the situations which might be met in the research laboratories. The non-specialist is even more likely not to be conversant with most modern techniques and reagents.

Nevertheless, it is perfectly possible for both the non-specialist and the graduate student beginning research in organic chemistry to carry out such reactions with success, provided that the appropriate precautions are taken and the proper experimental protocol is observed.

Much of this is common sense, given a knowledge of the properties of the reagents being used, as most general techniques are relatively straightforward. However, it is often very difficult for the beginner or non-specialist to find the appropriate information.

At Salford, we found ourselves handing out to students beginning research in organic chemistry a compilation of what we hoped was useful information on the practical aspects of organic synthesis, based on the authors' recent associations with some of the top synthetic organic research laboratories. We have gathered this information together in this book, and expanded it to cover some other areas, in the hope that it will be an aid to the specialist and non-specialist alike. Of course most research groups will have their own modifications and requirements, but on the whole the basic principles will remain the same.

This book is intended to be a guide to carrying out the types of reactions which are widely used in modern organic synthesis, and is concerned with basic technique. It is not intended to be a comprehensive survey of reagents and methods, but the appendix does contain some information on commonly used reagents.

If we have achieved our aims, users of this book should be able to approach their synthetic tasks with confidence. Organic synthesis is both exciting and satisfying, and provides opportunity for real creativity. If our book helps anyone along this particular path then our efforts will have been worthwhile.

CHAPTER 2

Safety

2.1 Safety is your primary responsibility

Chemical laboratories are potentially dangerous workplaces and accidents in the laboratory can have serious and tragic consequences. However, if you are aware of potential hazards, and work with due care and attention to safety, the risk of accidents is small. Some general guidelines for safety in the laboratory are presented in this section. In addition to these principles you *must* be familiar with the safety regulations in force in your area and the rules and guidelines applied by the administrators of your laboratory.

Your supervisor has a responsibility to warn you of the dangers associated with your work, and you should always consult him/her, or a safety officer, if you are unsure about potential hazards. However, your own safety, and that of your colleagues in the lab, is largely determined by *your* work practices. Always work carefully, use your common sense, and abide by the safety regulations.

Some important general principles of safe practice are summarized in the following rules

1. *Work carefully, do not take risks.*
   This covers basic rules such as always wearing safety spectacles, never working alone, and working neatly and unhurriedly.
2. *Assess the possible hazards before carrying out a reaction.*
   Find out about the dangers of handling unfamiliar chemicals or apparatus and take note of any necessary precautions.
3. *Know the accident and emergency procedures.*

It is vital to know what to do in case of an accident. This includes being familiar with the fire fighting and first aid equipment, and knowing how to get assistance from qualified personnel.

2.2 Safe working practice

It has been emphasized already that you should be familiar with the regulations and codes of practice pertaining in your laboratory. We will not discuss safety legislation here but some fundamental rules should be stressed. Never work alone in a laboratory. Always wear suitable safety spectacles and a cotton lab coat, and use other protection such as gloves, face masks, or safety shields if there is a particular hazard. Never eat, drink or smoke in a laboratory. Work at a safe, steady pace, and keep your bench and your lab clean and tidy. Familiarity breeds contempt; do not allow yourself to get careless with everyday dangers such as solvent flammability. Familiarize yourself with the location and operation of the safety equipment in your laboratory.

As regards specific hazards the chief rule is to carry out an assessment of the dangers involved before using an unfamiliar chemical or piece of apparatus. Some of the commonest hazards are described in the next section. Once you are aware of the possible dangers take all the necessary precautions, and ensure that you know what to do if an accident does occur.

Store your chemicals in clearly labelled containers, and abide by the regulations concerning storage of solvents and other hazardous materials. Dispose of waste chemicals safely, according to the approved procedures for your laboratory. Never pour organic compounds down the sink.

2.3 Common hazards

Always assess the risks involved *before* carrying out a reaction. Extensive compilations of information about the dangers posed by a large number of compounds are available (see Bibliography). Consult these references and your supervisor before using a compound or procedure which is new to you. In some areas safety legislation makes it mandatory to conduct such a safety audit, but even if it is not legally required, it should still be regarded as an essential preliminary before doing a reaction.

Remember to treat all compounds, especially new materials, with care. Avoid breathing vapours and do not allow solids or solutions to get on your skin. The majority of accidents are caused by a few common hazards. Some of the most frequently encountered dangers are listed in Table 2.1 and you should be aware of all of these, and always take appropriate precautions. Consult safety manuals and other Sections of this book for more information. Throughout the book safety warnings are highlighted in *bold italic* text.

Table 2.1

Common hazards in the chemical laboratory

Source	Hazard
Electrical equipment	Danger of fires caused by electrical sparks with solvent vapours, and a risk of electrocution with badly maintained equipment
Glassware	Danger of cuts, leaks of harmful compounds
Solvents	Most are extremely flammable
	Benzene, halogenated solvents are toxic
Vacuum apparatus	May implode violently
Pressure apparatus	May explode violently
Gas cylinders	May leak harmful gases or discharge violently (Chapter 7)
Strong acids	Extremely corrosive
	React violently with water, bases
	May produce harmful vapours
Strong bases	Extremely corrosive
	React violently with acids, protic solvents
Strong oxidizing agents	React violently with easily oxidizable compounds such as some organic solvents
Alkali metals	React violently with water, protic solvents and chlorinated solvents
Strong alkylating agents	Extremely toxic

In addition to these general warnings you should be aware of the severe hazards posed by some more specific families of compounds. The compounds listed in Table 2.2 pose a severe risk of explosion and those in Table 2.3 should be regarded as extremely toxic.

Table 2.2

Explosion hazards

Acetylene and metal acetylides

Alkali metals in contact with chlorinated solvents

Azides, both organic and inorganic

Diazo compounds and diazonium salts

Glass vacuum apparatus, e.g. Dewar flasks

Liquid oxygen and liquid air (formed by evaporation of liquid nitrogen)

Nitrates and polynitro compounds, e.g. TNT (trinitrotoluene)

Perchloric acid, perchlorates, and chlorates

Peroxides (formed in ethers and in alkenes on standing in air)

Table 2.3

Toxic and carcinogenic compounds

Compounds of heavy metals (arsenic, mercury, lead, selenium, thallium)

Alkylating agents including methyl iodide, dimethyl sulphate (*carcinogenic*)

Fluorine, chlorine and bromine

Hydrofluoric acid and metal fluorides

Cyanides and hydrogen cyanide

Oxalic acid and its salts, oxalyl chloride

Aromatic amines and nitro compounds

Ozone

Hydrogen sulphide

Phosgene

Osmium tetroxide

Benzene, polycyclic aromatics (*carcinogenic*)

Hexamethylphosphoric triamide (HMPA) (*carcinogenic*)

2.4 Accident and emergency procedures

Regrettably accidents are still all too common so it is vital that you know what to do if an accident does occur. You must be familiar with the fire fighting equipment in your lab (fire extinguishers, fire blankets, sand buckets) and you must know the procedures for summoning the fire brigade and for evacuating the building. In the case of injuries or exposure to harmful chemicals you should know who to summon to administer first aid, and how to get medical assistance. It is particularly important to know how to get help outside of normal working hours. If you are using particularly dangerous materials (such as cyanides) or equipment (such as high pressure apparatus) you should know about the relevant emergency procedures and take precautions such as having antidotes, protective equipment, or qualified personnel at hand. In the aftermath of an accident it is very important that you complete the required accident report forms, and take steps to avoid any possibility of a repeat.

Ask yourself now: are you familiar with accident procedures? If not you should *not* be working in the lab.

2.5 Bibliography

Guide to Safe Practices in Chemical Laboratories, Royal Society of Chemistry, 1986.
Hazards in the Chemical Laboratory, 5th ed., G.G. Luxon Ed., Royal Society of Chemistry, 1992.
The Sigma-Aldrich Library of Chemical Safety Data, 2nd ed., R.E. Lenga Ed., Sigma-Aldrich Corp., Milwaukee, 1987.
Dangerous Properties of Industrial Materials, 7th ed., N.I. Sax and R.J. Lewis, Van Nostrand Reinhold Co., New York, 1988.
First Aid Manual for Chemical Accidents, M.J. Lefevre, Dowden, Hutchinson and Ross, Stroudsburg, 1980.
Safe Storage of Laboratory Chemicals, D.A. Pipitone, Wiley, New York, 1984.
Handbook of Laboratory Waste Disposal, M.J. Pitt and E. Pitt, Wiley, New York, 1985.
The BDH (Merck) Hazard Data Sheets, Merck Ltd, 1990.

CHAPTER 3

Keeping Records of Laboratory Work

3.1 Introduction

No matter how high the standard of experimental technique employed during a reaction, the results will be of little use unless an accurate record is kept of how that reaction was carried out and of the data obtained on the product(s). Individuals or individual research groups will develop their own style for recording experimental data, but no matter what format you choose to follow, there are certain pieces of vital information which should always be included. In this section a format for keeping records of experimental data will be suggested and although this need not be strictly adhered to, it will be used to point out the essential features which should be included. It is suggested that records of experimental work and experimental data be kept in two complementary forms: The *lab notebook* should be a diary of experiments performed and should contain exact details of how experiments were carried out; A *data book* or set of *data sheets* should also be kept to record the physical data and preferred experimental procedure for each individual compound which has been synthesized.

3.2 The laboratory notebook

3.2.1 Why keep a lab book ?

Before any practical work is undertaken in the laboratory a sturdy hard-backed lab notebook should be obtained and a standard format for keeping the notebook should be decided upon. A good deal of thought should go into the layout of the lab book. It should be stressed that a lab book is not a format for polished report writing, but a daily log of work carried out in the lab. Some of the main reasons for keeping a lab book are:

1. In order that the exact procedure followed for a reaction can be referred to later. This can be very important even if the reaction was not successful. For instance, after several attempts to bring about a reaction have failed, it is often possible to review what has been done then carry out a more successful experiment.

2. It should be the main index point that will enable you to find experimental, literature and spectroscopic data on any compound which you have synthesized.

3. It is the main source of reference when you come to write reports, papers, theses etc.

4. It is a chronological diary of the experiments carried out and thus it should allow you to say exactly when a particular experiment was carried out.

5. In order that another worker can follow your work, it is very important to use a lab book style which is easily understood by others.

3.2.2 How to write a lab book

One of the most important points about keeping a lab book is that it is kept on the bench and written up as you perform the experiments. *It is bad practice to keep rough notes about experiments, then transfer the details to a lab book later.* This can cause many problems, for instance: the original notes can be lost; even with the strongest will, the exact truth often becomes distorted in transferring information to the lab book and small facts which may at the time seem unimportant are left out; it is also very easy to forget to rewrite an experiment altogether, especially if the reaction failed, and this can lead to much time wasting later. It is more important that the lab book be an accurate record of the way an experiment was performed, than for it to be in your neatest writing, although of course it should be legible.

An example of a format that is effective for general synthetic chemistry is outlined on page 11. This can be adjusted to personal needs but its essential features, which are listed below, should be included in any format chosen.

3.2.3 Suggested notebook format (Fig. 3.1)

1. *General layout*

It is good practice to start each new experiment on the next free right hand page of the notebook. This makes finding any particular experiment easier.

2. *Experiment number*

The experiment number is in the top right hand corner of the page and this is very important since it is used to reference all the compounds which are prepared. If a notebook with numbered pages is used, it is common for the experiment number to be the number of the page on which the experiment starts. The way in which the notebook is indexed is open to personal preference. In this system a researcher's first book is book *A*, then *B,C,D* etc. Fig. 3.1 therefore shows experiment *23* of book *A*. Compounds isolated from this experiment all carry the number *A23*, prefixed with the researchers initials (in this case *BB*). When more than one product is isolated from a reaction a suffix, *a, b, c* etc. is added to the reference number, *a* being the spot running highest on tlc, *b* the next, and so on. Thus, for this experiment two products were isolated and these carry reference numbers *BB A23a* and *BB A23b*. Using this system the origin of any synthetic sample can be determined very quickly.

3. *The date*

It is important that the date is always included.

4. *A reaction scheme indicating the proposed transformation*

This is always included at the top of the page so that an individual experiment is easily found. If the reaction proceeded as desired the scheme is left intact, but if the desired product was not obtained it can be crossed through in red to indicate this. If other products were also obtained they can be added, again in a different colour ink if desired. Thus, simply flicking through the lab book, looking at the schemes, can quickly provide a good deal of information. Some people prefer to write only the left-hand side of the equation until the experiment is complete.

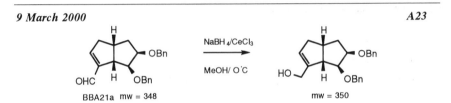

Ref., J.-L. Luche, L. Rodriguez-Hahn and P. Crabbe, J. Chem. Soc., Chem. Commun., 1978, 601

Substance	Quant.	Mol. wt.	m.moles	Equiv.	Source
BB A21a	200mg	348	0.57		p. A21
NaBH$_4$	27mg	38	0.71	2.84(H$^-$)	Aldrich
CeCl$_3$(0.4M/MeOH)	2ml		0.8	1.4	
MeOH	25ml				

Method:
        The aldehyde (200mg) and CeCl$_3$ solution (2ml) in MeOH (25ml), was cooled to 0°C, then treated with NaBH$_4$ (27mg) in MeOH (8ml).

TLC

10 min	30 min

4:1 Pet./EtOAc   4:1 Pet./EtOAc

← BB A23a (red/brown - vanilin)

← BB A23b (brown - vanilin)

SM   Rxn   SM   Rxn

After 30 min tlc shows no SM, but two products. MeOH was evaporated, CH$_2$Cl$_2$ (30ml) added and the mixture washed with 10% HCl (10ml) followed by satd. NaHCO$_3$ (3 x 10ml), dried and evaporated. (210mg crude)

Flash chromatography using 9:1 (pet. ether/EtOAc) on 8g of silica provided:

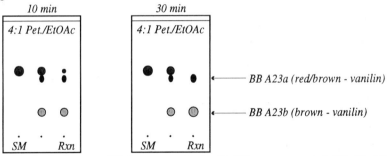

BB A23a 27mg (12% yield)- NMR (BB28), MS (BB19), IR (BB27), Data sheet 6 - looks like:

C$_{25}$H$_{30}$O$_4$
mw = 394

BB A23b 140mg (69% yield) - NMR (BB29), MS (BB20), IR (BB28), Data sheet 5 - OK for:

C$_{23}$H$_{26}$O$_3$
mw = 350

Comment:
        Next time use aqueous solvent - may avoid acetal formation

Figure 3.1

5.   *Literature references (if there are any)*

6.   *Quantities*

     The quantities of each ingredient of the reaction are listed at the beginning, together with the molecular weight, number of moles and number of molar equivalents. It is very useful to have this information available at a glance. Having the molecular weights on hand saves a good deal of time when going on to other reactions and when looking at mass spectra etc., but the real importance of this section is that the values contained here are critical when evaluating the outcome of a reaction, and you might want to adjust them in subsequent modifications of the procedure.

7.   *The procedure*

     This should be an exact account of the practical procedure carried out, including any spillages or other mishaps. It can be quite brief and not necessarily of publication standard, as long as it is understandable.

8.   *Reaction monitoring*

     Tlc is the most widely used method for reaction monitoring and it is very important to include a full size representation of the tlc plate(s) used, *giving the development solvent* and the stain used for visualization. Tlc gives us a true feel for the reaction and a good picture of a tlc is worth many words when it comes to following a procedure. In some cases hplc, gc, or other technique will be used to monitor the progression of the reaction and again a representation of this should be included. Some people find it convenient to draw the tlcs on the adjacent notebook page. (For more details about tlc, see Chapter 9)

9.   *Details of work-up and purification of the product(s)*

     For chromatography it is important to include the quantity and type of adsorbent, and the solvent system used for elution. Some people also like to include a tlc representation of the column fractions. If the product is purified by crystallization, record the solvent used and the m.p. If it is distilled describe the type of distillation set-up and record the b.p. and pressure.

10. *Cross references to the spectra and data book(sheet)*

All compounds should be given reference numbers, as described above, and these should be cross referenced with data book entries and the reference numbers of the corresponding spectra. Indeed, many people like to use the same number for the spectra. The yield of each compound isolated should also be given and if possible its structure.

11. *Finally, include any concluding remarks about the reaction*

3.3 Keeping records of data

When it comes to writing reports, papers, and especially theses, one of the most time consuming and tedious jobs is collecting together experimental and spectroscopic data for compounds, and it can be very frustrating to find that a particular piece of data has been mislaid or was never obtained. Also, if data collecting is left until the time of report writing, errors can easily creep into spectral assignments. It is a much better practice to collect data and make spectral assignments as your work progresses and keep this information stored in a standard format.

Whenever a significant compound has been synthesized a data sheet or data book page should be created for it. The format of a data sheet can vary according to personal preferences, but it should contain at least the following pieces of information:

1. The structure and molecular formula of the compound
2. An experimental procedure for the preparation of the compound, preferably in a style suitable for publication
3. An appropriate range of spectroscopic and chromatographic data which is sufficient to characterize the compound. *Full assignments of spectra should be entered in the data sheets as soon as the information is obtained, then when it comes to report writing most of the information required is on hand*
4. Cross references to spectra and lab notebook
5. Literature references, if there are any

3.3.1 *Purity, structure determination and characterization*

On preparing any compound for the first time, a competent organic chemist should always undertake a three step procedure:

Purification

• The compound must be isolated to a high state of purity, free of by-products and solvents.

Structure determination

• The structure of the compound must be established beyond any reasonable doubt, including stereochemisty, geometry, etc.

Characterization

• A range of data must be collected which will not only convince the wider chemical community of the structure and purity of the compound, but will also serve as that compound's identity.

It is worth considering these three separate processes very carefully, so that whenever you prepare a compound you will always collect the appropriate data for it. The skills of purification, structure determination and characterization are the mark of a competent organic chemist and you should strive to become as proficient as possible in each of them.

3.3.2   What types of data should be collected ?

It should be clear from the section above that you need to collect a range of data on each compound you prepare which will establish its purity, allow you to determine its structure, and act to identify it.

When you are determining the structure of a compound it is essential to scrutinize each piece of evidence carefully and critically until you are convinced that *all* the data is in accord with your proposed structure. Sometimes it will be quite easy to establish the structure of a compound beyond reasonable doubt, from say a simple nmr spectrum alone, especially if you know how the substance was synthesized. In other cases structure determination may be a major project in itself. In particular it is often a challenge to establish the absolute and/or relative stereochemical arrangement in chiral compounds. This may, for example, involve lengthy nmr decoupling and/or 2-D experiments to determine coupling constants and nOe experiments to determine through space interactions. It may be that a derivative of the original compound has to be made in order to obtain some crucial structural evidence. In some cases you may even have to resort to X-ray analysis.

No matter how simple or lengthy the task of structure determination, you need to collect a full range of spectral data, and some evidence of purity in order to convince the chemical community that you have characterized a 'new' compound. The range of data required for any compound you synthesize will depend on several factors, but most important of these are:

- *Is the compound new, or already known in the literature?*

- *Where is the work to be reported ?*

Before you start to collect data for a compound you should take these factors into account. For a known compound you will need at least enough data to match with that which is already published, but it is always best to collect enough evidence to be certain. If you have made a compound for the first time or by a new route you must characterize the compound fully, by collecting a *comprehensive* range of spectral data. The data must be in full accord with the proposed structure and be sufficient to convince any organic chemist that the correct assignment has been made. You must also have some evidence of purity. If you are working on a synthetic sequence where the structures of some of your intermediates are well established, it is not always necessary to treat each individual compound as a complete unknown. This is especially true if the reaction which has been carried out is a straightforward one.

Once you have decided that you need to characterize a compound, you should decide what data is required, bearing in mind where the work is to be reported. If there is the possibility of a patent application then combustion analysis is almost essential, together with a reasonable range of spectroscopic evidence to establish the structure beyond reasonable doubt. If you aim to publish work in the chemical literature and/or thesis form, you should agree with your supervisor on the standard which is to be adopted. Each journal has its own specific requirements not only for the type of data required, but also for the format in which it is to be presented. These specifications are usually published in the first issue of the year. Once you have decided which specification to comply with, make sure that you collect an adequate range of data for all your compounds *as you prepare them*. It will also save you and/or your supervisor many hours of work if you compile data in the format you have chosen *from the outset of your work*.

For most purposes the data specified below will normally be required for characterization purposes.

M.p. or b.p.

For solids the melting point should be specified, together with the solvent from which the compound was crystallized. If there is a distinctive crystal form and/or colour it is also worth specifying this also. For liquids the boiling range should be specified and the pressure.

Typical reporting formats are:

m.p. 56 - 57°C (from MeOH)

b.p. 120 -122°C at 5mmHg.

Molecular formula determination:

Data which establishes the compound's molecular formula is required. Traditionally an accurate combustion analysis (within 0.3 - 0.5%) has been used to determine the empirical formula of a compound, and also to justify that the compound is of 'high' chemical purity. However, combustion analysis data will be identical for structural isomers of any type (geometrical isomers, diastereoisomers, enantiomers etc.) and other spectroscopic or chromatographic methods will therefore be required in order to determine levels of isomeric impurities.

A typical reporting format is:

Found: C, 54.5; H, 5.8; N, 2.5%. $C_{25}H_{31}NO_{13}$ requires C, 54.2; H, 5.65; N, 2.5%

The molecular formula of a molecule can also be defined by high resolution mass spectrometry (hrms). The observed mass for the molecular ion or pseudo molecular ion must normally be within 5ppm of the calculated mass for EI (electron impact) measurements, or within 10ppm for CI (chemical ionization) measurements. It is important to note that high resolution mass spectrometry confirms that *some* molecules of a particular molecular formula are present in the sample, but does not give any indication of purity. Some other evidence of compound purity will therefore be required.

A typical reporting format is:

Found $[M+H]^+$ 230.1393. $C_{11}H_{20}O_4N$ requires 230.1391

¹H nmr spectrum

¹H nmr data is usually the most important piece of characterization data and you should try to analyse the spectrum as fully as possible - the values for every individual chemical shift and every coupling constant should be identified if possible. Proton nmr data is the most complex to report and it is important that you record it in an appropriate format as soon as you have interpreted the spectrum. A clean high field nmr spectrum can be used to justify purity and isomeric homogeneity and some journals now ask for a copy of the spectrum as supplementary material to a publication.

A typical reporting format is as follows:

- The nucleus should be specified as a subscript to the δ symbol.
- Instrument frequency, solvent and chemical shift standard should be given, e.g. δ_H (300 MHz, $CDCl_3$) - the standard can be in a preamble.
- Chemical shifts of individual signals and multiplets should be specified in sequence starting from the low δ end of the spectrum.
- Each chemical shift should be followed by a set of parentheses containing the following information, separated by comas, *in this order*

  i) number of nuclei under the signal, e.g. 3 H

  ii) multiplicity - s, d, t, q, dd etc. (br can be used for a broad signal)

  iii) coupling constants, e.g. $J_{1,2}$ 8, $J_{1,6}$ 3

  iv) an assignment of the proton, in the form CH_3CH_2 or 6-H

Thus a typical entry might begin:

δ_H (300 MHz, $CDCl_3$) 1.84 (1H, ddd, $J_{6a,6b}$ 12.5 $J_{6a,7}$ 6.5 $J_{6a,5}$ 1.5, 6a-H) 2.37 (1H, ddd, $J_{6a,6b}$ 12.5 $J_{6b,7}$ 11.0 $J_{6b,5}$ 9.0, 6b-H) 2.68 (1H, m, 5-H), 3.89 (4 H, br s, OCH_2CH_2O), ...etc.

Ir spectrum

Do not forget to record the ir spectrum for *all* compounds and record the frequencies of the main bands.

Presented in the form:

$\nu_{max}/\ cm^{-1}$ 2935 (CH), 1742 (C=O), 1200 (C-O).

Low resolution mass spectrum

It is preferable to have a mass spectrum which shows the molecular ion of the compound and for this reason soft ionisation techniques such as chemical ionization (CI), fast atom bombardment (FAB) and electrospray are now widely used instead of or as well as electron impact (EI) ionization.

It is important to report the conditions under which the spectrum was run together with a list of the main peaks. It is also useful to work out the structures of the fragment ions.

Presented in the form:

m/z (NH_3, Cl) 230 ([M+NH_4]$^+$ 100%), 213 ([M + H]$^+$, 11), 91 (50)

Optical activity data

If your compound is optically active the specific rotation should be measured. If it is a known compound it is best to record the rotation in the same solvent and at a similar concentration to that reported previously. The solvent, concentration and temperature should always be reported!

Presented in the form:

$[\alpha]_D^{20}$ +89 (c 1.25 in $CHCl_3$)

Note that optical rotations are sometimes measured at frequencies other than that of the sodium-D line.

Other data

The types of data given above should be considered as a *minimum* requirement for characterising a compound, but several other types of data are useful.

^{13}C nmr spectrum

A fully characterised ^{13}C nmr spectrum is always a useful piece of data and can also be used as evidence of purity. A ^{13}C spectrum is particularly useful when a compound has most of its proton resonances in the same region of the spectrum. The level of assignment will depend on how well the spectrum can be interpreted. It may simply be a list of chemical shifts, it may be a list of chemical shifts with signals assigned as CH_3, CH_2, CH or C, or you may be able to give actual assignments to individual carbon atoms.

Presented in the form:

δ_C (75 MHz, $CDCl_3$) 20.1 (CH_3), 44.6 (CH_2), 46.7 (CH),............

Specialised ^1H nmr data

There is a variety of powerful modern nmr techniques which can be of great assistance in structure determination - nOe effects; 2-D nmr etc. The data from such techniques should be reported as appropriate.

Chromatography data

Hplc and or gc data may be reported to indicate purity or isomer ratios present in a mixture. Remember to record the conditions and type of column used as well as retention times.

Evidence of enantiomeric purity

If you have any data which defines the enantiomeric purity of the compound (chiral hplc, nmr with chiral shift reagent, etc.) this should be presented.

There are many other types of data which can be collected and reported in appropriate circumstances, uv, ord (optical rotatory dispersion), cd (circular dichroism) etc.

Once you have a clear idea of the range of data you need to collect for your compounds, you should establish a format for recording it. The formats suggested below can be adapted to comply with the specification of any publisher.

3.3.3 Formats for data records

If you choose to keep a data book, it is a good idea to start each new entry on the next free right hand page of the book. The data for some compounds will not fill the space allocated, but it is best to allow a reasonable space, because some compounds will require a large amount of spectroscopic data for characterization.

Data sheets are an alternative to a data book. These can be of a standardized design, as shown in Fig. 3.2, with spaces for each type of data to be filled in. The advantage of this system is that it is easy to see at a glance whether a particular piece of data has been obtained for a compound. However, two disadvantages of the fixed format data sheet are: it does not provide the flexibility which is often required to record diverse types of data used to characterize any particular compound; and it tends to encourage the mistaken idea that every compound requires the same characterization data.

A system we prefer for data records is to have a computerized standard blank data sheet. The grid shown in Fig. 3.2 can be used, but a more simple and flexible format is shown in Fig. 3.3. Either of these can easily be constructed in a word-processing package.

When a new compound is synthesized a blank data sheet is modified with the data of the compound and the data record is tailored to the needs of that particular compound. Any of the data types which are inappropriate for characterization of a particular compound can be removed, and any additional data types can be added. The data blank shown is designed so that data can be added in the exact style required for publication or for experimental entries for theses (it can easily be adjusted to other formats). An example of a completed data record is shown in Fig. 3.4 (see Section 3.4.3 for details of how this type of data record can be transformed into an experimental section entry).

Once the data sheets have been printed out they are kept in a ring file, to make a very flexible data book. Similar non-computerized record systems can also be devised but the advantage of using a computer is that the record can easily be updated at any time. If a researcher is conscientious about keeping these data records up to date, much of the tedious hard work is done when it comes to writing reports or theses.

Data Sheet

Chem. Name	
Lit. Refs.	
Scheme	
Method	

Notebook Ref		Spectra Refs.	
Mol. Formula		m.p./b.p.	
Tlc - R_f (solv)		$[\alpha]_D$ (c, solv.)	
ν_{max}/cm^{-1}			

^{1}H NMR	δ value	No. H	Mult.	j value/Hz (coupled proton)	Proton

^{13}C NMR	
m/z	
Anal./HRMS	

Figure 3.2

DATA SHEET

Chemical Name:

Scheme:

Lit. Refs:

Method:

Mol. Formula:
Notebook ref.:
Spectra refs.:
m.p./b.p. $^{\circ}$C
Tlc : R_f (light petroleum/ethyl acetate,   )
$[\alpha]_D^{20}$: $^{\circ}$ (c =   , CHCl$_3$)
v_{max}/cm^{-1}
δ_H(MHz,   ):
δ_C (MHz,   ):
λ_{max}/nm
m/z (NH$_4$, CI):
Found: M$^+$ C H N O requires
Analysis: Found: C %; H %; N %
                    C H N O requires C %; H %; N %

Figure 3.3

(±)-*exo*,*exo*-7,8-Dibenzyloxy-*exo*-*cis*-bicyclo[3.3.0]-oct-2-en-4-ol-2-carboxaldehyde

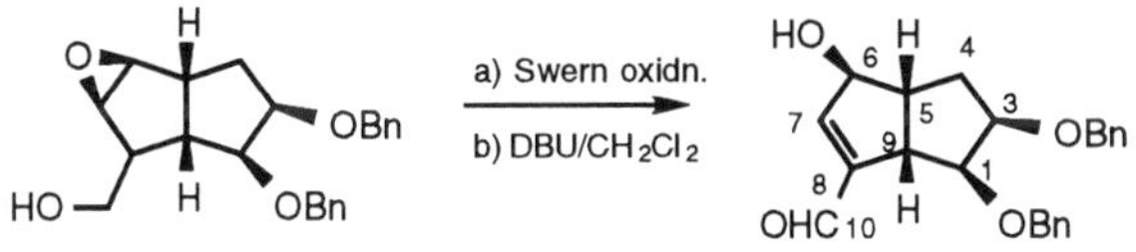

Method:

To a stirred solution of oxalyl chloride (91 cm^3, 1.0 mmol) in dry dichloromethane (15 cm^3) at -78°C was added, a solution of dimethylsulphoxide (182 cm^3, 2.35 mmol) in dichloromethane (1 cm^3). After 5 min *exo*,*exo*-7,8-dibenzyloxy-*exo*-3,4-epoxy-*exo*-2-(hydroxymethyl)-*cis*-bicyclo[3.3.0]-octane (200 mg, 0.55 mmol) in dichloromethane (2 cm^3) was added dropwise and after a further 20 min triethylamine (1 cm^3, 7.61 mmol) was added. After 10 min at -78°C the mixture was allowed to warm to room temperature, then partitioned between 2M hydrochloric acid (30 cm^3) and dichloromethane (2 x 30 cm^3). The organic extract was washed with sat. aq. sodium hydrogen carbonate (40 cm^3), dried (MgSO$_4$) and the solvent evaporated off. A solution of the crude product and 1,5-diazabicyclo[5.4.0]undec-5-ene (167 mg, 1.1 mmol), in dichloromethane (15 cm^3) was stirred at room temperature for 2 h. The solution was then poured into 2M hydrochloric acid (20 cm^3) and extracted with dichloromethane (2 x 30 cm^3). The organic extract was washed with sat. aq. sodium hydrogen carbonate (40 cm^3), dried (MgSO$_4$) and the solvent evaporated off. Purification by flash chromatography [light petroleum/ethyl acetate (1:1)] provided the title compound, (159 mg, 80%) as a colourless oil

Mol. Formula:	C$_{23}$H$_{24}$O$_4$ (mw = 364)
Notebook:	BB D45b
Tlc:	R$_f$ 0.23 (uv-active; light pet./ethyl acetate, 1:1)
ν$_{max}$/cm^{-1}	3400 (OH), 3075, 3050, 2750 (CHO),1690 (C=O)

δ$_H$ (300 MHz, CDCl$_3$):

δ value	no. H	Mult.	J value/Hz (coupled proton)	proton(s)
1.84	1H	ddd	$J_{6a,6b}$ 12.5 $J_{6a,7}$ 6.5 $J_{6a,5}$ 1.5	6a-H
2.37	1H	ddd	$J_{6a,6b}$ 12.5 $J_{6b,7}$ 11.0 $J_{6b,5}$ 9.0	6b-H
2.68	1H	m	$J_{5,6b}$ 9.0 $J_{1,5}$ 8.0 $J_{5,6a}$ 1.5 $J_{4,5}$ 1.5	5-H
3.47	1H	ddd	$J_{6b,7}$ 11.0, $J_{6a,7}$ 6.5, $J_{7,8}$ 4.5	7-H
3.58	1H	ddd	$J_{1,5}$ 8.0, $J_{1,3}$ 1.0, $J_{1,8}$ 1.0	1-H
3.80	1H	dd	$J_{7,8}$ 4.5, $J_{1,8}$ 1.0	8-H
4.28	1H	d	J 11	CH$_2$Ph
4.38	1H	d	J 12	CH$_2$Ph
4.50	1H	dd	$J_{4,5}$ 1.5, $J_{3,4}$ 1.0	4-H
4.68	1H	d	J 12	CH$_2$Ph
4.77	1H	d	J 11	CH$_2$Ph
6.55	1H	~t	$J_{1,3}$ 1.0, $J_{3,4}$ 1.0	3-H
7.1-7.5	10H	m		Ar-**H**
9.73	1H	s		CHO

m/z (+ve FAB, thioglycerol):	365 ([M + H]$^+$, 10%), 364 (M$^+$, 7), 57 (100)
Found:	[M + H]$^+$ 365.1751. C$_{23}$H$_{24}$O$_4$ requires 365.1753

Figure 3.4

3.4 Some tips on report and thesis preparation

Most research projects will culminate in the submission of a report or thesis and the preparation of such documents can be quite daunting if you are inexperienced in such matters. In this section we will try to provide some general guidance, paying particular attention to presentation of experimental results. The backbone of any thesis or report in organic chemistry is the body of experimental results that have been gathered during the project. For this reason, the results should be reviewed and evaluated before starting to write the report.

3.4.1   Sections of a report or thesis

Most detailed organic chemistry reports will consist of three main sections:

The Introduction

In this section the project is introduced, the origins of the project and the original objectives are outlined, and all relevant background work is reviewed and referenced.

The Discussion

This is a detailed description of the work that was actually carried out. It should be a guided tour of the project presenting the reader with a realistic picture of how the project developed and the discoveries that were made during the course of the work. There should be discussions of objectives that were concluded successfully and explanations of those which failed.

The Experimental Section

This contains the method of preparation for each compound together with a set of data which is adequate to characterize it. The range of data presented and the style of presentation must conform to strict technical requirements and it is most important that the appropriate standards are understood and adhered to.

3.4.2   Planning a report or thesis

Planning is the key to writing a good report or thesis. From the outset you should aim to construct the document so that it has a logical structure and is easy to follow. The mark of a good report is that it will be a useful document for someone *less* knowledgeable than yourself, who may for example take over the project. A thesis that only your supervisor can follow will be of little use since he or she is presumably familiar with the results already. Always remember when you write a report that *you are the expert* and your aim is to produce a document which will be a useful resource for other workers.

Planning the Experimental Section

If you have kept good data records for your compounds, as suggested in Section 3.3, you will have a good basis for starting your thesis plan. First of all carefully review the original objectives of the project and then organize your compound data sheets into a logical order based on how the project progressed with respect to the objectives. This collection of data sheets will eventually become your experimental section.

Planning the Discussion

Having reviewed your experimental work you should have a clear picture of the key achievements made during the course of the project. During many research projects the original objectives will not have been accomplished, and it may at first seem that much of the work has been unsuccessful. Nevertheless, it is most unusual if significant discoveries have not been made. You should construct your report so that achievements and discoveries are highlighted. Try to take an overview of your work so that you can partition it into separate topics if necessary. A large body of work, such as that contained in a thesis is normally best broken down into Sections or Chapters of related work. Try to be logical about pieces of work that are grouped into Chapters and arrange them and the Sections within them in logical sequences. For example, a target synthesis project could be broken down into different Chapters for different approaches to the molecule or different Chapters for approaches to different fragments of the molecule. There could be several ways of sectionalizing your results. Try to find the best way of constructing the document so that a reader can follow a logical path through your results. Also, try to section the report so that a reader can easily find and interpret any individual topic. Once you have assessed your work in this way, you should have a working structure plan for your report. Do not treat this as a final plan and do not be afraid to rearrange Sections or Chapters as the report takes form.

Planning the Introduction

Having reviewed your own work, you should be in a good position to appreciate what background material should be covered in the introduction. The nature of the introduction will depend on the type of project. For example if the aim of the project was to synthesize a natural product it would be appropriate to report the origin of the compound, its structure determination, properties, biosynthesis (if known), and then review any other synthetic approaches that have been reported. Your own synthetic strategy could then be explained as a lead-in to your discussion section.

By the time you come to write a thesis or report you should be familiar with the background to the project and you will probably have a large volume of background reference material. Before you start any further serious literature searching make sure you are familiar with any previous reports relating to the project and it is a good idea to discuss the background to the project with your supervisor. Then you will have to get down to some serious library work, searching for any relevant papers and reviews which you have not already seen. See Chapter 17 for more information on literature searching.

The next stage is to create an outline plan for your review. Decide on the background material you wish to cover and group together papers that are related to one another. Groups of related papers will form the Sections of your review. When arranging the material into a sequence it is good practice to start from a broad base of background work and gradually focus in on material which is of direct relevance to your project. Try to order the material logically. Think back to when you started work on the project and try to construct the type of document that you would have ideally liked to have been given at that time. Once you have grouped together related pieces of background material and arranged the topics into a logical order, you will have an outline plan to start writing from.

3.4.3   Writing the report or thesis

Once you have a good thesis plan you should be in a position to start writing effectively. Remember to break down the material into Sections, the Sections into Sub-sections and if necessary the Sub-sections into Sub-sub-sections, which will make it much easier for the reader to find particular pieces of information. Remember that very few people will ever want to read a scientific report from cover to cover; they will usually want to look up a particular section, so try to make them self-contained. The hierarchical sectioning system used in this book is now widely used and can be recommended.

Writing the Introduction

Before you start writing you should have copies of background reports and papers grouped together. It is a good idea to keep each group in a separate folder. Make sure you are fully conversant with each piece of work before you write about it and, when you are reviewing the work of others, be careful that you put the results in their appropriate context. Now you have the material in manageable packages you can start writing.

- *Start with the broader issues first, outlining the general area of chemistry into which the project belongs, then gradually focus in on material that is directly related to the project.*
- *Prepare reaction schemes which clearly illustrate each piece of work that you review. Good schemes are generally the most important aspect of a review.*
- *Be as concise as possible with the text, using it primarily to explain the information contained in the Schemes.*
- *Describe the objectives of studies that have been carried out previously and summarize the results, but do not give lengthy details of experimental work.*

As each background document is covered add it to a reference list *in the proper format* with all the authors names and initials, the year, volume, page and correct abbreviated journal name. It will save you so much time if you get this information detailed correctly from the outset. The format used for presenting references must be consistent and should conform to a recognized style (as required by a journal or your institute).

When you are writing the first draft of the document do not attempt to finalize the numbering of references, schemes or structures, but you do need to identify them. One way to do this is to number the structures and references within sections, the items in the section you write first being numbered 1, 2, 3....., those in the second being numbered 101, 102, 103..... etc. If there are extra references or structures to be inserted at any stage they can be inserted as 26a, 26b, etc. as appropriate. At this stage it is simply important that the numbers in the text match those on the schemes and in the list of references.

As you cover the topics you will most probably want to arrange some sections in a different order than in your original plan and you may also decide that some extra ground has to be covered and/or that some of the originally planned material can be cut. Before you worry too much about the final ordering try to write all the topics and make sure each is covered accurately. After this most of the technical work is done, but a good deal of work is normally required before the review is readable. First of all organize the topics into their final arrangement within Sections, Sub-sections etc. Then make sure that there is a flow through the document from one section to the next. At this stage your introduction may be quite bland and read like a list of reactions. There are several simple things you can do to make it more interesting for the reader.

- *Make sure each new topic is introduced so that the subject matter is put into some sort of context.*
- *Compare and contrast results from different research groups.*
- *Highlight what you consider to be the major achievements in the field prior to your study and explain why you think they are important.*
- *Try to set the stage for your own study.*
- *Finally outline the objectives of your work clearly and with respect to previous studies.*

When you are happy with the content and coverage of your introduction let someone else read it, preferably your supervisor. This other person may offer a lot of criticism of your first draft, but do not be put off, in fact you should welcome constructive suggestions. If you have time, it is best to put the introduction on one side for a while before redrafting and editing, then read it all the way through carefully. Try to iron out any remaining mistakes as you read through then take an overview. The final editing is crucial.

- *Make sure you have not missed out any important points.*
- *Make sure there is a logical structure and a good flow from one topic to the next.*
- *Make sure the topics are balanced, trimming any that are too expansive and enhancing any that are too brief.*

- *Do not be afraid to edit material that is not relevant and make all your writing as concise as possible.*
- *Make sure you are satisfied that you have provided a good introduction to your project for anyone new to the field of study - this the crucial test for an introduction!*

When you have made all the final changes, you can then start at the beginning of the document numbering all the Schemes, the compounds and the references in sequence. Then read and check through the document several times carefully, correcting mistakes and making final minor changes.

Writing the Discussion

Much of the advice given above for writing the introduction is true for the discussion section also, but certain distinct differences in style are required when reporting your own results. Once you have an outline plan you can start writing your results within sections. Give your writing some structure and avoid sections that are simply repetitive lists of the experiments carried out. The following sequence is a useful way to describe work carried out within a topic or for describing an individual experiment.

i) *Always start by outlining the objectives of the topic or the particular experiment you are about to describe.*

ii) *Outline the work or experiment that was carried out, pointing out any special features of the experiment(s). Here you do need to provide more detail than for the experiments described in your introduction.*

iii) *Report the outcome of the experiment(s) and how you determined the outcome. This may be simple or it may involve a detailed discussion of your interpretation of complex spectral data.*

iv) *Avoid bland statements, such as 'the reaction failed' or 'the reaction gave 50% of the required product.'*

v) *Draw conclusions from the outcome of the experiments and compare this to the original objectives.*

vi) *In the light of the outcome, outline follow up studies*

This is basically a cyclic six step process, step vi) often leads back into step i) for a new experiment or topic. There will almost certainly be significant portions of your work which did not proceed according to the original objectives. However, it is rare that there is nothing to report concerning these 'failed experiments.' Indeed, it is often necessary to discuss the actual outcome of such reactions in some detail before proceeding to introduce alternative strategies that were adopted to solve the problem. Occasionally results from a reaction that did not go according to plan are more interesting than the predicted outcome would have been and in such cases a good deal of discussion will be required. Pay attention to the following points throughout the discussion:

- *Draw reaction schemes to illustrate your work clearly and make sure that they are located near to the part of the text that refers to them. Avoid making the reader search for a numbered structure many pages from where it is described in the text, and if necessary repeat the structure.*
- *Make good use of tables to compare results from related studies.*
- *Try to present your studies in a positive manner.*
- *Make connections between the work covered under different sections.*
- *Where you have interesting results to report highlight them and explain clearly the significance of your discoveries.*
- *Compare and contrast your results with other published work, especially contemporary work.*
- *Modulate your writing style and do not be afraid to point out when results are interesting, curious, confusing, incompatible with those in the literature etc.*

The final stages of editing and redrafting the Discussion are as described for the Introduction above.

Writing the Experimental Section

A bland, repetitive presentation style could be considered an asset in the experimental section, so very little imagination is required here, but it is technically more demanding than the other sections. This section of a thesis or dissertation can be measured against rigid standards and a badly prepared experimental section is likely to be a fatal flaw!

The following guidelines provide some tips on how to write a good experimental section which will be acceptable to a thesis examiner or journal referee. The recommendations in this section are provided so that preparation of that crucial manuscript will be a straightforward task and not the nightmare encountered by many graduate students upon writing-up.

- *Start on your FIRST DAY in the laboratory by collecting good data sets for all your compounds as your work progresses.*

- *In your experimental section be absolutely consistent about the style in which your data is presented and stick to an agreed journal or institutional standard.*
- *Make sure you have a comprehensive selection of data for all new compounds, including proof of molecular formula and of purity. Do not forget individual pieces of data that are easily overlooked, such as ir data and the specific rotation for optically active compounds.*
- *Present your data to an accuracy within the limits of experimental error UNDER WHICH IT WAS COLLECTED - not according to the nominal tolerance limits of the instrument or simply as given on an electronic print out! For example δ values from routine high field ^1H nmr spectra should not be quoted to an accuracy of more than 0.01ppm and coupling constants should not be quoted to an accuracy of more than 0.5Hz. ^{13}C nmr δ values should not be quoted to more than one decimal place, positions of ir bands should be given in whole wave numbers, m.p.s and b.p.s should be give to the nearest degree, etc. Apart from anything else data becomes less easy to decipher, and is therefore LESS useful when quoted to higher tolerance limits.*

At the beginning of the experimental section there should be a preamble describing the instrumentation used to record the data and the conditions that the data was recorded under. Be careful to check the make and model numbers of instruments and be sure you record conditions accurately. It is useful to state standard units used in the reported data, as this will save you repeating these units in each data set. It is also useful to record specifications for standard methods used, for example the type of silica and tlc plates used for chromatography, the drying agent used for routine drying of organic solutions, etc. Again this will save a lot of duplication in each entry. An example of a typical experimental preamble is shown in Fig. 3.5 and this pattern can easily be modified to your needs.

Melting point determinations were carried out on a KÖffler block electrothermal apparatus and were recorded uncorrected. Infra-red absorption spectra were run either neat (for liquids) or as nujol mulls (for solids) on a Perkin-Elmer 1710 FT-IR instrument. [1]H nmr spectra were recorded at 300 MHz on a Bruker AC-300 instrument, as solutions in deuterochloroform, unless stated otherwise. Chemical shifts are referenced to tetramethylsilane and J values are rounded to the nearest 0.5 Hz. Mass spectra were recorded at low resolution on a Finnigan 4500 instrument and at high resolution on a Kratos Concept 1-S instrument, under electron impact (EI) conditions, or chemical ionisation (CI) using ammonia, as specified. After aqueous work-up of reaction mixtures, organic solutions were routinely dried with anhydrous magnesium sulphate and 'evaporation' or 'evaporated' refer to removal of solvent on a rotary evaporator. Thin layer chromatography was carried out using Merck Kieselgel 60 F_{254} glass backed plates. The plates were visualized by the use of a UV lamp, or by dipping in a solution of vanillin in ethanolic sulphuric acid, followed by heating. Silica gel 60 (particle sizes 40-63 μ) supplied by E.M. Merck was employed for flash chromatography.

Figure 3.5

For preparations of organic compounds, the title of each data entry should usually be the IUPAC[1] name of the compound prepared, although a generic name, such as 'Ketone **21**', is sometimes acceptable. This should be followed by a description of the method by which the compound was prepared. A more or less standard format has developed for reporting such methods. The method must be precise, but should be concise. Repetitious statements should be avoided. For example, once it has been stated that the reaction mixture was stirred, or under nitrogen, or at -78°C, these facts need not be restated unless the conditions are altered. This is the one part of any report where all the individual write-ups should be presented in the same style. It is therefore worth learning a standard style and using it for all your preparative methods. If you had the forethought to record the preparative method for each compound you prepared on a data sheet at the time of preparation (as outlined in Section 3.3), this can be used directly in your

1. IUPAC Handbook on naming structures.

Experimental Section, otherwise you will have to write out a new method for each entry, which can be very tedious and time consuming.

A standard format for an experimental method of preparation is broken down in the table below:

To a stirred solution of *[compound A]* (*X* g, *Y* mmol), at *Z*°C, in *[solvent Q]* (*A* cm^3), *[reagent P]* (*R* g, *S* mmol) was added over *T* min.	*This part describes addition of reagent(s) to substrates - must include amounts in g and moles.*
After *D* h *[aqueous reagent U]* (*H* cm^3) was added, the organic layer was separated and washed twice more with *[aqueous reagent U]* (*H* cm^3), then dried over magnesium sulphate and the solvent evaporated off.	*This part says how long the reaction was left, how it was worked-up and how the product was isolated.*
The *[...coloured solid/liquid]* residue was purified by *[flash chromatography (solvent system) or distillation, b.p. L°C at W mmHg or recrystallization, m.p. T°C]*, to provide the title compound as a *[....coloured solid/liquid] M* g, *XX*% yield,	*This says how the product was purified - choose the appropriate statement and insert the required information.*

This standard format can be used, with slight modifications, for many different methods by simply changing the variables which are in italics and adding any other important statements.

The method of preparation should be followed by a list of characterization data as outlined in Section 3.3.2.

If computerized data sheets have been used (as recommended in Section 3.3.3), experimental entries can be created easily. Simply by deleting headings, the data sheet shown in Fig. 3.4 is transformed into a tabulated experimental entry, as used in some reports or theses, shown in Fig. 3.6. Alternatively, some formatting can also be removed to provide a journal style experimental entry as shown in Fig. 3.7.

(±)-*exo*,*exo*-7,8-Dibenzyloxy-*exo*-*cis*-bicyclo[3.3.0]-oct-2-en-4-ol-2-carboxaldehyde

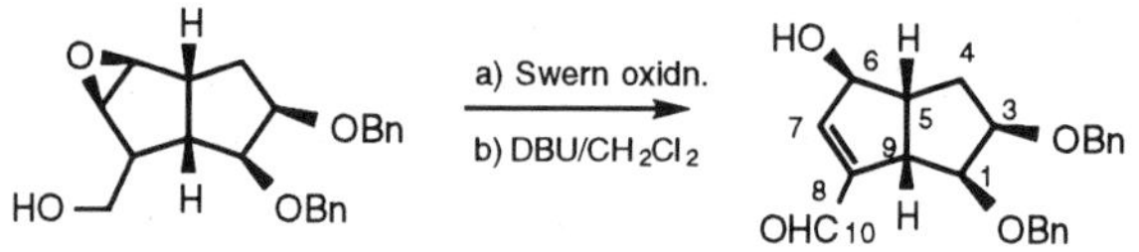

To a stirred solution of oxalyl chloride (91 cm^3, 1.0 mmol) in dry dichloromethane (15 cm^3) at -78°C was added a solution of dimethylsulphoxide (182 cm^3, 2.35 mmol) in dichloromethane (1 cm^3). After 5 min *exo*,*exo*-7,8-dibenzyloxy-*exo*-3,4-epoxy-*exo*-2-(hydroxymethyl)-*cis*-bicyclo[3.3.0]-octane (200 mg, 0.55 mmol) in dichloromethane (2 cm^3) was added dropwise and after a further 20 min triethylamine (1 cm^3, 7.61 mmol) was added. After 10 min at -78°C the mixture was allowed to warm to room temperature, then partitioned between 2M hydrochloric acid (30 cm^3) and dichloromethane (2 x 30 cm^3). The organic extract was washed with sat. aq. sodium hydrogen carbonate (40 cm^3), dried (MgSO$_4$) and the solvent evaporated off. A solution of the crude product and 1,5-diazabicyclo[5.4.0]undec-5-ene (167 mg, 1.1 mmol) in dichloromethane (15 cm^3) was stirred at room temperature for 2 h. The solution was then poured into 2M hydrochloric acid (20 cm^3) and extracted with dichloromethane (2 x 30 cm^3). The organic extract was washed with sat. aq. sodium hydrogen carbonate (40 cm^3), dried (MgSO$_4$) and the solvent evaporated off. Purification by flash chromatography [light petroleum/ethyl acetate (1:1)] provided the title compound, (159 mg, 80%) as a colourless oil

Tlc:	R$_f$ 0.23 (uv-active; light pet./ethyl acetate, 1:1)
v_{max}/cm^{-1}	3400 (OH), 3075, 3050, 2750 (CHO), 1690 (C=O)

δ$_H$ (300 MHz, CDCl$_3$):

δ value	no. H	Mult.	J value/Hz (coupled proton)	proton(s)
1.84	1H	ddd	$J_{6a,6b}$ 12.5 $J_{6a,7}$ 6.5 $J_{6a,5}$ 1.5	6a-H
2.37	1H	ddd	$J_{6a,6b}$ 12.5 $J_{6b,7}$ 11.0 $J_{6b,5}$ 9.0	6b-H
2.68	1H	m	$J_{5,6b}$ 9.0 $J_{1,5}$ 8.0 $J_{5,6a}$ 1.5 $J_{4,5}$ 1.5	5-H
3.47	1H	ddd	$J_{6b,7}$ 11.0, $J_{6a,7}$ 6.5, $J_{7,8}$ 4.5	7-H
3.58	1H	ddd	$J_{1,5}$ 8.0, $J_{1,3}$ 1.0, $J_{1,8}$ 1.0	1-H
3.80	1H	dd	$J_{7,8}$ 4.5, $J_{1,8}$ 1.0	8-H
4.28	1H	d	J 11	CH$_2$Ph
4.38	1H	d	J 12	CH$_2$Ph
4.50	1H	dd	$J_{4,5}$ 1.5, $J_{3,4}$ 1.0	4-H
4.68	1H	d	J 12	CH$_2$Ph
4.77	1H	d	J 11	CH$_2$Ph
6.55	1H	~t	$J_{1,3}$ 1.0, $J_{3,4}$ 1.0	3-H
7.1-7.5	10H	m		Ar-**H**
9.73	1H	s		CHO

m/z (+ve FAB, thioglycerol):	365 ([M + H]$^+$, 10%), 364 (M$^+$, 7), 57 (100)
Found:	[M + H]$^+$ 365.1751. C$_{23}$H$_{24}$O$_4$ requires 365.1753

Figure 3.6

(±)-exo,exo-7,8-Dibenzyloxy-exo-cis-bicyclo[3.3.0]-oct-2-en-4-ol-2-carboxaldehyde -
To a stirred solution of oxalyl chloride (91 cm^3, 1.0 mmol) in dry dichloromethane
(15 cm^3) at -78°C was added a solution of dimethylsulphoxide (182 cm^3, 2.35 mmol)
in dichloromethane (1 cm^3). After 5 min *exo,exo*-7,8-dibenzyloxy-*exo*-3,4-epoxy-*exo*-
2-(hydroxymethyl)-*cis*-bicyclo[3.3.0]-octane (200 mg, 0.55 mmol) in dichloromethane
(2 cm^3) was added dropwise and after a further 20 min triethylamine (1 cm^3, 7.61
mmol) was added. After 10 min at -78°C the mixture was allowed to warm to room
temperature, then partitioned between 2M hydrochloric acid (30 cm^3) and
dichloromethane (2 x 30 cm^3). The organic extract was washed with sat. aq. sodium
hydrogen carbonate (40 cm^3), dried (MgSO$_4$) and the solvent evaporated off. A
solution of the crude product and 1,5-diazabicyclo[5.4.0]undec-5-ene (167 mg, 1.1
mmol) in dichloromethane (15 cm^3) was stirred at room temperature for 2 h. The
solution was then poured into 2M hydrochloric acid (20 cm^3) and extracted with
dichloromethane (2 x 30 cm^3). The organic extract was washed with sat. aq. sodium
hydrogen carbonate (40 cm^3), dried (MgSO$_4$) and the solvent evaporated off.
Purification by flash chromatography [light petroleum/ethyl acetate (1:1)] provided
the title compound, (159 mg, 80%) as a colourless oil, v_{max}/cm^{-1} 3400 (OH), 3075,
3050, 2750 (CHO),1690 (C=O); δ_H (300 MHz, CDCl$_3$) 1.84 (1 H, ddd $J_{6a,6b}$ 12.5
$J_{6a,7}$ 6.5 $J_{6a,5}$ 1.5, 6a-H), 2.37 (1 H, ddd $J_{6a,6b}$ 12.5 $J_{6b,7}$ 11.0 $J_{6b,5}$ 9.0, 6b-H), 2.68
(1 H, m $J_{5,6b}$ 9.0 $J_{1,5}$ 8.0 $J_{5,6a}$ 1.5 $J_{4,5}$ 1.5, 5-H), 3.47 (1 H, ddd $J_{6b,7}$ 11.0, $J_{6a,7}$ 6.5,
$J_{7,8}$ 4.5, 7-H), 3.58 (1 H, ddd $J_{1,5}$ 8.0, $J_{1,3}$ 1.0, $J_{1,8}$ 1.0, 1-H), 3.80 (1 H, dd $J_{7,8}$ 4.5,
$J_{1,8}$ 1.0, 8-H), 4.28 (1 H, d J 11, CH$_2$Ph), 4.38 (1 H, d J 12, CH$_2$Ph), 4.50 (1 H, dd
$J_{4,5}$ 1.5, $J_{3,4}$ 1.0, 4-H), 4.68 (1 H, d J 12, CH$_2$Ph), 4.77 (1 H, d J 11, CH$_2$Ph), 6.55
(1 H, ~t $J_{1,3}$ 1.0, $J_{3,4}$ 1.0, 3-H), 7.1-7.5 (10 H, m, Ar-H), 9.73 (1H, s, CHO); *m/z*
(+ve FAB, thioglycerol) 365 ([M + H]$^+$, 10%), 364 (M$^+$, 7), 57 (100), (Found: [M +
H]$^+$ 365.1751. C$_{23}$H$_{24}$O$_4$ requires 365.1753

Figure 3.7

General References:

Instructions for authors section of major journals which carry full
papers, for example *J. Chem. Soc., Perkin 1* or *J. Am. Chem. Soc.* Such
instructions are usually found in the first issue of each year. Note that
the styles differ slightly from one journal to another.

CHAPTER 4

Equipping the Laboratory and the Bench

4.1 Introduction

In this chapter we will describe general laboratory and bench equipment which we have found to be of use when working with modern organic chemical reactions.

Many modern procedures which have now become standard methodology in organic chemistry require dry reaction conditions, and often an inert atmosphere. This has had a dramatic effect on the way efficient laboratory facilities are arranged. Not so long ago reactions involving anhydrous, inert conditions were rare and it was expedient to arrange the equipment for such procedures on a one-off basis. However, now that this type of reaction is common place, it makes sense to set up the laboratory in such a way that reactions under inert conditions can be carried out as a matter of routine. This chapter is written with this principle in mind. Much of the equipment introduced here will be discussed in more detail in subsequent chapters.

4.2 Setting up the laboratory

The basic furniture provided in organic chemistry laboratories will vary considerably from one establishment to another and clearly any advice given in this chapter will have to be tailored to the facilities available. The ideal layout of the lab is also a very subjective matter and the advice given here is therefore not intended to be taken as gospel, but simply reflects the experiences of the authors from various laboratories in which they have worked.

When setting up the lab it is usual for some areas of bench space to be set aside for communal apparatus, and other parts to be allocated as individual benches. Clearly, the areas which are assigned as communal bench space will depend on the amount and type of communal equipment which is to be installed. In this chapter some pieces of equipment will be identified as communal and others as part of the individual bench kit, but the classification will vary from one laboratory to another. The distinction will to some extent depend on the type of work being undertaken, but will also depend on the budget and space available.

Unless alternative office space is provided, it is a good idea to have some desk space in the lab where workers can read and write, away from areas used for chemicals. Desk space may also be required for small computers which are a common feature of the modern organic chemistry lab. Drawers or filing cabinets are useful for the safe storage of spectra and other paperwork, and a blackboard is an invaluable laboratory aid.

The area which constitutes an individual 'bench' will vary considerably from one lab to another. In our view all procedures involving organic chemicals should be carried out in an efficient fume cupboard. This implies that each full-time worker in an organic chemistry lab permanently requires *at least* one metre of fume cupboard space. However, in practice it is often the case, particularly in academic labs, that much less fume cupboard space is actually provided, and fume cupboards may be communal. In this chapter the term 'bench' will refer to the space occupied by an individual worker and it will be assumed that this space incorporates an adequate area of fume cupboard, *where all reactions are carried out*.

4.3 General laboratory equipment

In this section we describe the communal facilities that should be found in a laboratory in which modern organic chemistry is undertaken. Careful thought should go into the placing of communal equipment. It may be quite reasonable to place unattended items such as a refrigerator or oven in an awkward corner, but a piece of apparatus at which a person will be working should be in a position where the operator has enough room to work without hindering anybody else. Equipment such as stills, which are used regularly by all members of the lab, should be located so that they are easily accessible, without causing a disturbance to anyone working close by.

Rotary evaporators

Rotary evaporators are perhaps the most heavily used pieces of equipment in an organic research lab. They should therefore be located conveniently throughout the lab, or preferably, each worker should have his or her own on the bench.

Refrigerator and/or freezer

A refrigerator and/or freezer should be provided to store chemicals, and this should never be used for food storage.

Glass drying oven(s)

Again these should be conveniently located.

Vacuum oven

This may be used infrequently, but is still a very valuable piece of equipment.

Balances

A two place balance is indispensable and a four place balance needs to be available for small scale work. When locating these remember that they are used frequently, by people carrying out careful work, so do not put them in the way of others; also avoid drafts and vibrations.

Kugelrohr bulb to bulb distillation apparatus (e.g. Buchi)

This apparatus is invaluable when preparing relatively small quantities of high-boiling liquids. It is relatively mobile and can be moved from bench to bench and attached to various vacuum sources. However, it is useful to locate it near to a high vacuum pump, where it is normally used.

Vacuum pumps

Vacuum at various levels will be required around the lab.

A modest vacuum source (about 20 - 50mmHg) should be available at each bench and is sufficient for rotary evaporation of most solvents, filtering under vacuum, distillation of relatively volatile oils, and similar tasks. This level of vacuum is sometimes provided by a house vacuum system. Alternatively, a water aspirator or small diaphragm pump can be used.

A reliable medium vacuum (about 12mm Hg) can be provided by oil-free Teflon lined diaphragm pumps (e.g. KNF, Vacuubrand, Divac). These pumps are extremely useful and can often be used instead of either water aspirators or oil pumps. They circumvent the suck-back problems associated with water aspirators and have the advantage over oil pumps that they are solvent resistant and can be used without liquid nitrogen trapping systems. They can be used for rotary evaporation, vacuum distillation and many other general tasks. For most purposes they can also be used as the vacuum source for double manifold type inert gas lines (described in Section 4.4.3, Chapter 5 and Chapter 8)

High vacuum oil pumps (providing <1mm Hg) are necessary for certain operations and these include removing last traces of solvent from small quantities of product and distillation of high-boiling oils. They can also be used as the vacuum source for double manifold type inert gas lines. A good two stage rotary pump will provide a vacuum of <0.1mm and this is adequate for almost any task in an organic chemistry lab. It is useful to have a general purpose communal pump, attached to a single manifold (Fig. 4.1), which can be used for removing trace amounts of solvent, and for distillations. Ideally a good quality pump should be reserved for distillations, so that a reliable high vacuum can always be obtained. This pump should be fitted with an efficient trapping system (see Chapter 8) to avoid solvent contamination, and can be either on a mobile trolley or in a fixed position, depending on the bench and floor space available. A pump which is used for distillations only does not require a manifold to be attached.

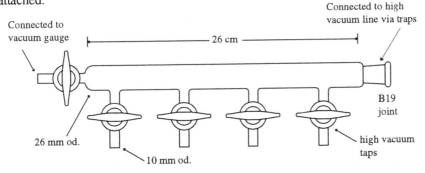

Figure 4.1

If sufficient resources are available each bench should have a medium or high vacuum pump, connected to a double manifold, but in many labs vacuum pumps are considered to be communal equipment (see Chapter 8 for more details about vacuum pumps).

Inert gases

A constant supply of dry inert gas (usually argon or nitrogen) is now essential in an organic chemistry lab. Some labs will have nitrogen or argon on tap and this is ideal, provided there are enough outlets and the gas is dry. However, in most labs gas is supplied from cylinders, which take up a lot of space and are quite expensive to rent. It is inefficient in a modern lab if there are too few inert gas outlets, and it is therefore important when setting up the lab to decide how many gas outlets are required. We suggest that there should be at least one outlet for each person working in the lab, plus one for each piece of apparatus (such as a still) which is kept permanently under inert gas. This may suggest that the same number of cylinders is required, but the cost of this is usually prohibitive and space can also be a problem. One efficient way to use cylinders to service a large number of permanent inert gas lines, is to fit them with multi-way needle valves (e.g. 3-way). These are available from gas line hardware suppliers at a very modest cost and give several gas outlets for the price of one!

Having decided on the number of gas cylinders required these should be positioned in the laboratory so that semi-permanent PVC (Tygon) tubing lines can easily be taken to each bench and to each piece of apparatus requiring a supply. Some of the apparatus commonly used in conjunction with inert gases will be described below and in Chapters 7 and 9.

Solvent stills

Two basic types of solvent still are found in the lab. One is for distillation of solvents for routine use and the other is for distillation of ultra-dry solvents for carrying out reactions under dry conditions.

If solvents are purchased in bulk (drums) they will normally need to be redistilled even for routine laboratory use. Large stills (about 5 litres) will normally be required for this purpose. Typical solvents requiring such stills are petroleum ether, ethyl acetate, dichloromethane and other solvents according to particular needs. There are many problems associated with routine distillation of all lab solvents: stills take up valuable space; large stills are hazardous; the process is time consuming; redistilled solvents are not always available when needed; there is always a considerable volume of solvent wasted; and disposing of these residues costs money. We have found that it is generally more efficient and cost effective to purchase bottled solvents that are sufficiently pure for routine use without distillation. We find that GPR grade solvents from a reputable dealer are adequate for most purposes, including large and medium scale chromatography. Small quantities of solvents for special purposes can easily be redistilled using compact stills and some typical designs for these are given in Chapter 5.

Some solvents, such as tetrahydrofuran and diethyl ether (there may be others you require also) are frequently required in very dry form, and it is inconvenient to set up a still each time a small quantity is required for a trial reaction. The efficiency of an organic lab is therefore enhanced by having these dry solvents available 'on-tap' from permanent stills, which provide very effective drying, and are kept under an inert atmosphere. Modern designs for permanent stills used for such distillations are very compact, and do not cause inconvenience (see Chapter 5). For every solvent which is required in dry form regularly a permanent still should be installed in the lab. It is also worth keeping a spare distillation apparatus on hand for occasions when a less common dry solvent is required.

General distillation equipment

Apart from standard Quickfit equipment, labs should also have some one-piece distillation kits, which provide more effective distillation. A short-path distillation apparatus is very useful for low-hold-up, high-throughput distillations, particularly on a small scale. These can now be purchased but are very expensive. They can easily be constructed by a glassblower according to the design shown in Fig. 4.2. Note that distance B on the diagram is crucial and corresponds to the length of a Quickfit thermometer from bulb to joint.

For fractional distillation a one-piece apparatus incorporating small fractionating columns is useful, and again two sizes are frequently used (Fig. 4.3; see also Chapters 6 and 11).

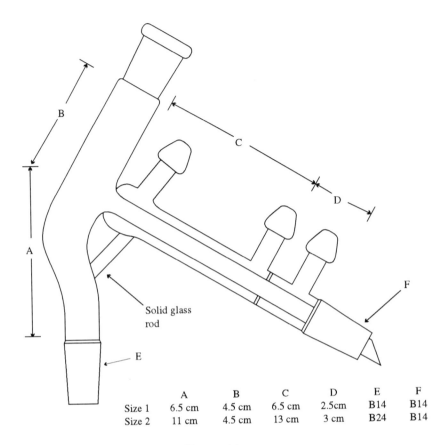

	A	B	C	D	E	F
Size 1	6.5 cm	4.5 cm	6.5 cm	2.5cm	B14	B14
Size 2	11 cm	4.5 cm	13 cm	3 cm	B24	B14

Figure 4.2

General laboratory glassware

In any lab there will be quite a number of pieces of glassware which it is only practical to keep communally; this is especially true for large equipment. However, the nature of this equipment will vary a good deal, depending on the type of work being undertaken.

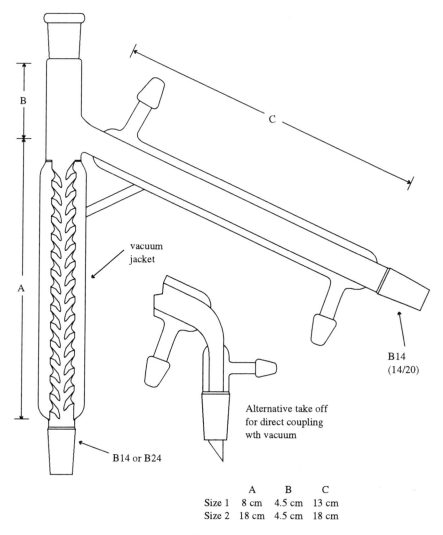

B

A

vacuum
jacket

B14
(14/20)

Alternative take off
for direct coupling
wth vacuum

B14 or B24

C

	A	B	C
Size 1	8 cm	4.5 cm	13 cm
Size 2	18 cm	4.5 cm	18 cm

Figure 4.3

Chromatographic equipment

Tlc is by far the most widely used technique for rapid, routine reaction monitoring and it is essential that the lab should have permanent facilities for visualization of tlc plates. A range of solvent dips, a heat gun, an iodine tank and a small uv lamp are usually all that is necessary (see Chapter 9).

More sophisticated chromatographic techniques are also finding much wider application in organic chemistry labs, and both hplc and capillary gc instruments are now quite common-place. These techniques complement tlc as analytical techniques and, although an initial investment of time is necessary to learn how to use the instrumentation, this pays good dividends in the long run (see Chapters 9 and 11).

4.4 The individual bench

There are a variety of factors that influence the proportion of laboratory equipment that is part of an individual bench set, and that which is communal. In most academic research labs workers find it most convenient to keep a more or less complete set of routine glassware as part of the bench kit. The individual then looks after the set, replaces broken equipment, and keeps it clean, so that any particular piece of equipment is available when required. On the other hand, in an industrial lab there is normally an extensive range of communal routine glassware which is washed by lab helpers, and there is little need for workers to keep a full set of bench equipment. In either type of situation there are certain pieces of equipment which individuals usually like to keep for personal use, including such things as syringes and chromatography columns. Whether more specialized or sophisticated equipment is communal or personal will depend, to some extent, on how frequently it is used, and on the funds available. A list of equipment which we find useful to keep as part of the bench set is given below, with a brief description of some of the more specialized items.

4.4.1 Routine glassware

A typical bench set of routine glassware, excluding specialized equipment, is given below.

Item	Quantity
Reduction adapter B14 socket B24 cone	1
Reduction adapter B10 socket B14 cone	1
Adapter tube B14 cone	1
Expansion adapter B24 socket B14 cone	1
Expansion adapter B29 socket B24 cone	1
Beakers 10ml	2
Beakers 25ml	4
Beakers 100ml	4
Beakers 400ml	2
Measuring Cylinders 10ml	1

Measuring Cylinders 25ml	1
Measuring Cylinders 100ml	1
Measuring Cylinders (stoppered) 1000ml	1
Filter funnels 15.00cm	1
Filter funnels 7.5cm	1
Separating funnels 1000ml	1
Separating funnels 250ml	1
Separating funnels 50ml	1
Separating funnels 25ml	1
Conical flask 100ml	4
Conical flask 250ml	2
Conical flask 500ml	2
Round-bottom flask 10ml B14 socket	6
Round-bottom flask 50ml B14 socket	6
Round-bottom flask 100ml B24 socket	6
Round-bottom flask 250ml B24 socket	3
Round-bottom flask 500ml B24 socket	2
Round-bottom flask 1000ml B24 socket	1

4.4.2 Personal items

There are certain items which individuals usually prefer to keep for their own personal use, even when routine glassware is communal. Here is a list of standard, commercially-available items which we think should be kept as the bench kit:

Item	Quantity
Spatulas (various sizes)	-
Magnetic followers 25.00mm	4
Magnetic followers 12.5mm	4
Magnetic followers 25.00mm (oval)	1
Magnetic followers 50.00mm (oval)	1
Syringe needles 6 inch S-S Luer fit	6
Syringe needles 12 inch S-S Luer fit	6
Micro syringe 100 μl	2
Syringe with metal Luer lock 1ml	2
Syringe with metal Luer lock 2ml	2
Syringe with metal Luer lock 5ml	2
Syringe with metal Luer lock 10ml	2
Luer lock tubing adapter	1
Thermometers -10 to 360°C	1
Thermometers -10 to 110°C	1
Thermometers -100 to 10°C	1
Thermometers, Quickfit B14 -10 to 360°C	1
Pasteur pipettes	1 box
Vials (1 & 5 dram)	1 box

4.4.3 Specialized personal items

There are some specialized pieces of glassware which we find to be extremely useful to have as part of the bench set and these will be briefly described here, with diagrams showing how they can be constructed by a glassblower. For full details of how they are used see the appropriate chapter.

The double manifold

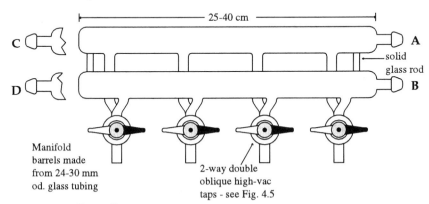

Connections:
A connected an inert gas cyclinder via a bubbler (unless bubbler attached at C)
B connected to a vacuum pump via a trapping system (see Chapter 8)
C (optional outlet) connected to bubbler
D (optional outlet) connected to vacuum guage

Figure 4.4

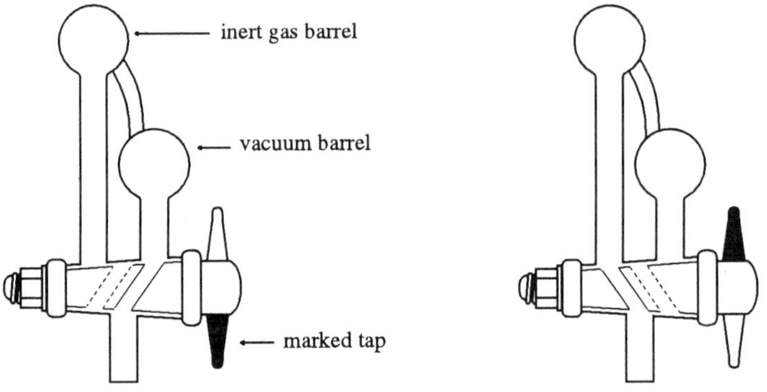

(a) tap switched to vacuum   (b) tap switched to inert gas

Figure 4.5 Cross section through a double oblique tap

If you wish to carry out reactions under dry and/or inert conditions as a matter of routine, the double manifold is perhaps the single most useful item of equipment that will enable you to do this. We recommend that this be a standard item of equipment, permanently fitted to every individual bench and connected to the laboratory inert gas supply by PVC (Tygon) tubing. The manifold consists of two glass barrels, one evacuated and one filled with an inert gas. An outlet is supplied by either barrel of the manifold *via* a two-way double oblique tap (see Figs. 4.4 and 4.5). Thus, at the turn of the tap, equipment connected to the manifold can be alternately evacuated or filled with inert gas.

A bubbler should be incorporated in the line from the cylinder which feeds the inert gas barrel and it should have a built in anti suck-back valve to avoid oil contaminating the manifold (Fig. 4.6).

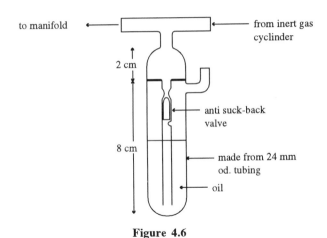

to manifold ←   ← from inert gas cyclinder

2 cm

anti suck-back valve

8 cm

made from 24 mm od. tubing

oil

Figure 4.6

A rotary pump is normally connected to the vacuum barrel of the manifold and a schematic diagram showing the complete set-up is shown in Fig. 4.7. For most purposes a Teflon lined diaphragm pump can be used and this avoids the necessity for solvent traps.

Although it is preferable for workers to have their own individual manifold, in some labs this may not be possible, because of insufficient fume cupboard space or lack of pumps. In this case, manifolds can be treated as communal items of equipment, but discipline must be observed to keep the manifold clean and avoid cross-contamination. For communal use larger manifolds with more outlets are generally used.

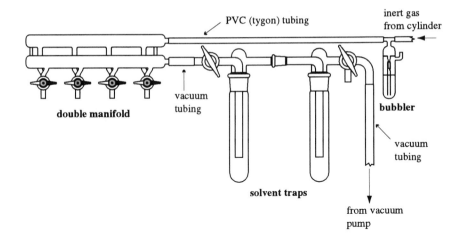

Figure 4.7

Another type of manifold which is useful for carrying out reactions under inert atmosphere is the 'spaghetti tubing' manifold (Fig. 4.8). This is a single barrel inert gas manifold, fitted with narrow bore Teflon tubing outlets. Each outlet has a syringe needle attached which can be pushed through a septum to provide an inert atmosphere within a flask. This type of system is particularly useful for small scale work.

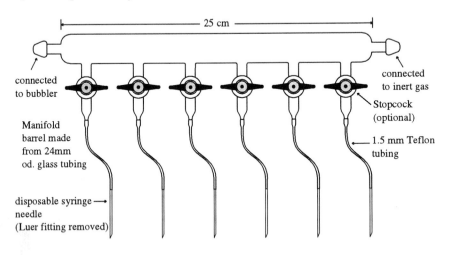

Figure 4.8

For full descriptions of how to use manifolds see Chapters 5 and 9.

Three-way Quickfit gas inlet T taps

A simple piece of equipment which is applicable to a variety of tasks is a 3-way Teflon tap connected to a Quickfit cone joint (Fig. 4.9).

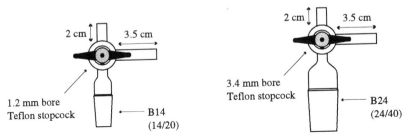

Figure 4.9

These are so universally useful that we suggest you have at least three of the B14 size and two of the B24 size as part of your personal bench set. They are particularly useful when used in conjunction with a double manifold. With the inert gas from the manifold connected to the horizontal inlet and the tap in position A (Fig. 4.10), a reaction flask can be kept under a slight positive inert gas pressure. If the gas flow is increased and the tap is turned to position B, liquids can be introduced *via* the vertical inlet, whilst maintaining an inert atmosphere (see Chapters 6 and 9 for more details). Another simple use is for connecting flasks to a high vacuum system for removal of last traces of solvent.

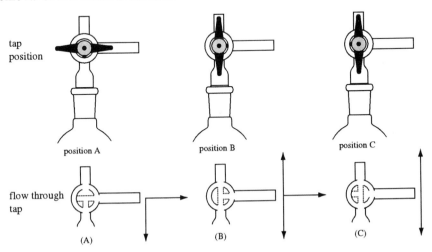

Figure 4.10

Filtration aids

Rapid filtration is often needed, and the speed of filtration is increased dramatically by applying pressure or a vacuum. There are two types of one-piece sintered funnels which we find very useful, as shown in Fig. 4.11. The parallel-sided type are constructed by fusing a circular sinter into the appropriate diameter glass tubing, which is then joined to a cone joint and a piece of narrow-bore tubing (about 10mm od). It is useful to have two or three of these in the bench kit, ranging in diameters of between 1 and 10cm. The larger ones are particularly valuable for filtering off drying agent, leaving the dried solution in a round-bottom flask, from which solvent can be evaporated directly. One or two of the smaller Hirsch-style funnels are useful and these are more commonly used for filtering off crystals after small-scale recrystallizations, the mother liquor being conveniently deposited in a round-bottom flask. These can be made starting from a sintered funnel, but a glassblower can make them more cheaply by inserting a circular sinter into a narrow tube, then forming the funnel from this.

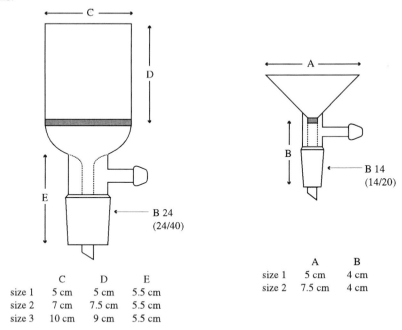

	C	D	E
size 1	5 cm	5 cm	5.5 cm
size 2	7 cm	7.5 cm	5.5 cm
size 3	10 cm	9 cm	5.5 cm

	A	B
size 1	5 cm	4 cm
size 2	7.5 cm	4 cm

Figure 4.11

A one-piece filtration apparatus, as shown in Fig. 4.12a is very useful for small scale filtration of solutions prior to recrystallization. It is often used in conjunction with a Craig tube (Fig. 4.12b, see Chapter 11 for more detail). A sintered funnel (Fig. 4.13) for filtration under inert atmosphere is also useful (see Chapter 6 for more details).

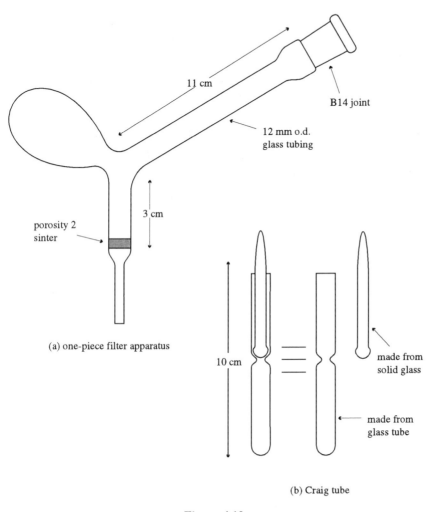

11 cm

B14 joint

12 mm o.d.
glass tubing

3 cm

porosity 2
sinter

(a) one-piece filter apparatus

10 cm

made from
solid glass

made from
glass tube

(b) Craig tube

Figure 4.12

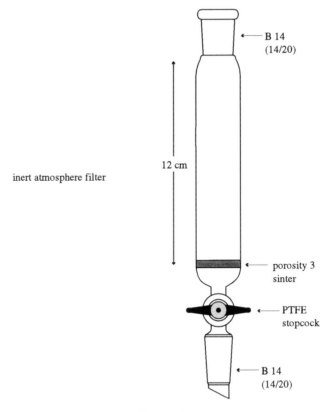

B 14
(14/20)

12 cm

inert atmosphere filter

porosity 3
sinter

PTFE
stopcock

B 14
(14/20)

Figure 4.13

Glassware for chromatography

Flash chromatography (see Chapter 11) is probably the most effective method available for rapid routine purification and separation of reaction products. To gain expertise in flash chromatography it is a sound idea to have a familiar set of columns on hand, and it is useful to have a set of about five, ranging in diameter from about 5mm to 50mm, as part of the bench kit (Fig. 4.14b). A convenient length for all columns is 25cm, except for those with a very narrow bore, which can be shorter. Reservoirs will also be required for flash chromatography, 250ml, 500ml, and 1 litre being useful sizes (Fig. 4.14c).' Equipment with ordinary glass joints, held together with rubber bands, can be used for flash chromatography but Rodaviss joints are much more reliable and safer to use. If you use a high pressure gas source to pressurize flash columns a useful item is a simple 'flash valve' (Fig. 4.14a) that will limit the pressure applied to the column, thus making the operation safer by reducing the possibility of column fracture. We find it safer and more convenient for each worker to have a small diaphragm pump to pressurize columns. The use of this equipment is described in detail in Chapter 11.

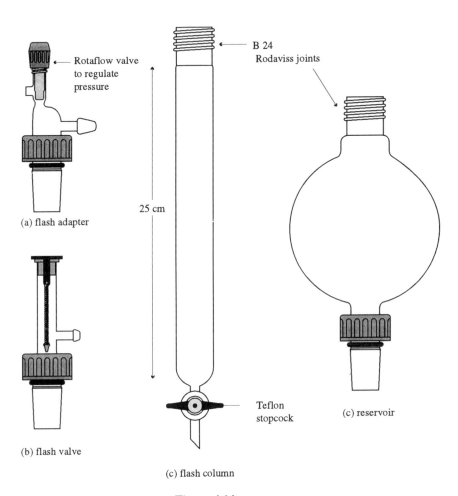

(a) flash adapter

(b) flash valve

25 cm

B 24
Rodaviss joints

Teflon
stopcock

(c) flash column

(c) reservoir

Figure 4.14

CHAPTER 5

Purification and Drying of Solvents

5.1 Introduction

The use of appropriately purified solvents is vital to the success of the procedures for which they are used. It is important to note that the degree of purity and dryness required depends on the intended application, so when choosing a method from this chapter remember to pick one which is appropriate for your purpose. Consult Appendix 1 for useful data about commonly used organic solvents.

Remember that solvents are hazardous materials and beware particularly of the flammability of the hydrocarbons and ethers, the toxicity of benzene, chloroform, and carbon tetrachloride, and the possibility of peroxide contamination of ethereal solvents.

5.2 Purification of solvents

The most commonly used grade of solvent is 'reagent grade' which typically means 97-99% purity with small amounts of water and other volatile impurities as listed in the specification. This is adequate for use in extractions and in many chemical reactions. However, some applications are more demanding and require either the use of commercial solvents of higher purity, or the purification of commercially available products. Notable examples include the following.

1. Reactions involving strongly basic organometallic compounds, e.g. Grignard reagents, organolithiums, and metal hydrides, require the use of carefully dried solvents. Anhydrous solvents (< 50ppm of water)

are available but they are more expensive and less reliable than solvents dried by the techniques described below.

2. Solvents used for spectroscopy, especially nmr and uv, should be of high purity. Many suppliers provide 'spectroscopic grade' solvents which are particularly suitable for uv spectroscopy because ultraviolet absorbing impurities have been removed.

3. Solvents used for chromatography should always be fractionally distilled to ensure that non-volatile impurities are removed. Solvents for hplc should be of high purity and again many suppliers provide special 'hplc grade' solvents, which have been purified and filtered to remove contaminants which might degrade hplc columns.

4. All applications involving quantitative analysis require the use of 'analytical grade' (typically >99.5% purity) solvents. It is good practice in general to use high quality solvents for the purification of your products, and it is particularly important to use pure solvents when purifying samples for microanalysis.

In most cases purification of a solvent simply involves drying and distilling it, so the next sections are devoted to drying agents and methods used for drying common solvents, and the final section contains descriptions of typical continuous stills. Apart from water the only other commonly encountered contaminant is peroxidic material formed by aerial oxidation of ethereal solvents. Methods for dealing with these dangerous impurities are described in the section on the purification of diethyl ether.

5.3 Drying agents

Drying agents fall into two broad categories, those used for preliminary drying and the drying of extracts, and those used for rigorous drying. The pre-drying agents are largely interchangeable with each other and the choice is usually limited only by the chemical reactivity of some of the reagents. Preliminary drying of solvents, prior to rigorous drying, is essential unless the solvent already has a low (< 0.1%) water content. Of necessity the reagents used for thorough drying are very reactive and they must be treated with great care. In particular, it is important to make the right choice of drying agent for the solvent in question, in order to avoid dangerous or undesirable reactions between the solvent and the drying agent, and care must be taken to ensure the safe destruction and disposal of excess reagent remaining in solvent residues.

When the solvent is to be distilled after standing over a desiccant, the drying agent should be filtered off before distillation if it removes water reversibly, e.g. by hydrate formation ($MgSO_4$, $CaCl_2$), or by absorption (molecular sieves). The solvent can be distilled without removal of the desiccant in cases where water removal is irreversible (CaH_2, P_2O_5).

The recommendations in this and the following section are largely based on the work of Burfield and Smithers *et al.* who have carried out quantitative studies on the efficiency of drying agents for a wide range of solvents.[1] Their work has supplanted earlier studies many of which are of doubtful reliability. Other useful sources of information include Perrin and Armarego,[2] and Riddick, Bunger, and Sakano.[3]

Alumina, Al_2O_3

Neutral or basic alumina of activity grade I is an efficient drying agent for hydrocarbons, 5% w/v loading giving extremely dry solvents. It is also useful for the purification of chloroform and removal of peroxides.

Barium oxide, BaO

The commercially available anhydrous product is an inexpensive drying agent which is useful for amines and pyridines (30-50ppm after standing for 24h over 5% w/v). It is strongly basic and is ineffective for alcohols and dipolar aprotic solvents.

Boric anhydride, B_2O_3

Recommended drying agent for acetone, and effective also for thorough drying of acetonitrile.

1.   (a) D.R. Burfield, K.H. Lee, and R.H. Smithers, *J. Org. Chem.*, 1977, **42**, 3060; (b) D.R. Burfield, G.H. Gan, and R.H. Smithers, *J. Appl. Chem. Biotechnol.*, 1978, **28**, 23; (c) D.R. Burfield and R.H. Smithers, *J. Org. Chem.*, 1978, **43**, 3966; (d) D.R. Burfield and R.H. Smithers, *J. Chem. Technol. Biotechnol.*, 1980, **30**, 491; (e) D.R. Burfield, R.H. Smithers, and A.S.C. Tan, *J. Org. Chem.*, 1981, **46**, 629; (f) D.R. Burfield and R.H. Smithers, *J. Chem. Educ.*, 1982, **59**, 703; (g) D.R. Burfield and R.H. Smithers, *J. Org. Chem.*, 1983, **48**, 2420.

2.   D.D. Perrin and W.L.F. Armarego, *Purification of Laboratory Chemicals*, 3rd ed., Pergamon, Oxford, 1988.

3.   J.A. Riddick, W.B. Bunger, and T.K. Sakano, *Organic Solvents, Physical Properties and Methods of Purification*, 4th ed., Wiley-Interscience, New York, 1986.

Calcium chloride, CaCl$_2$

Both the powder and pellet forms are effective for pre-drying hydrocarbons and ethers. It reacts with acids, alcohols, amines, and some carbonyl compounds.

Calcium hydride, CaH$_2$

The reagent of choice for rigorous drying amines, pyridines, and HMPA, and effective also for hydrocarbons, alcohols, ethers, and DMF. It is available in powdered or granular form; the granular is preferable if it is to be stored for any length of time. The granules should be crushed immediately before use and residues should be destroyed by *careful* addition of water (H$_2$ evolution).

Calcium sulphate, CaSO$_4$

Available as 'Drierite', it is only suitable for drying organic extracts. The blue self-indicating version should not be used to dry liquids because the coloured compound may leach into the solvent.

Lithium aluminium hydride, LiAlH$_4$

Although widely used for drying ethers, it is less effective than other methods and is extremely dangerous.
Its use is strongly discouraged.

Magnesium, Mg

Recommended for methanol and ethanol.

Magnesium sulphate, MgSO$_4$

The monohydrate is fast acting and has a high capacity (forms a heptahydrate), making it the desiccant of choice for organic extracts. It is slightly acidic so care is required with very sensitive compounds. It is not efficient enough to be useful for pre-drying.

Molecular sieves

These are sodium and calcium aluminosilicates which have cage-like crystal lattice structures containing pores of various sizes, depending on their constitution. They can absorb small molecules, such as water, which can fit into the pores. The most commonly used types 3A, 4A, and 5A have pore sizes of approximately 3Å, 4Å, and 5Å respectively, and they are available in bead or powder form. After activation at 250-320°C for a minimum of 3h they are probably the most powerful desiccants available.[1(b)] They can be stored in a desiccator or in an oven at >100°C for a few weeks but they are rapidly hydrated in the air. If you are doubtful about the effectiveness of an old batch, place a few beads in the palm of your hand and add a drop of water - if the sieves are active you should feel a distinctly exothermic reaction. In most cases extremely dry solvent can be obtained simply by batchwise drying over sieves, i.e. allowing the solvent to stand over 5% w/v of sieves for 12h, decanting, adding a second batch of sieves etc. Sieves absorb water reversibly so a solvent should always be decanted from the sieves prior to distillation. 4A Beads are recommended for thorough drying of amines, DMF, DMSO, and HMPA, and almost all rigorously dried solvents are best stored over 5% w/v of 4A sieves. However, only the 3A form is suitable for drying acetonitrile, methanol, and ethanol, and higher alcohols require the use of powdered 3A sieves. They are not useful for drying acetone because they cause self-condensation. Provided that they are not discoloured, sieves can be reused by washing well with a volatile organic solvent, allowing to dry, drying at 100°C for several hours, and then reactivating at 300°C.

Phosphorus pentoxide, P_2O_5

(Causes burns) Although a rapid and efficient desiccant its use is limited by its high chemical reactivity. It reacts with alcohols, amines, acids, and carbonyl compounds and causes significant decomposition of HMPA, DMSO, and acetone. It is useful for drying acetonitrile and may be used for hydrocarbons and ethers but is less convenient than other reagents. It is often used in desiccators. It is best decomposed by careful portionwise addition to ice-water followed by neutralization with base (do not add water to P_2O_5, the mixture may become so hot that the vessel could crack). It is extremely efficient for drying gases and is available in a convenient form, mixed with an inert support, so that it does not become syrupy.

Potassium hydroxide, KOH

(Causes burns) Freshly powdered KOH is a good drying agent for amines and pyridines but is inferior to calcium hydride. It should not be used with base sensitive solvents.

Sodium, Na

Sodium is widely used to dry hydrocarbons and ethers. It may be formed into wire using a sodium press or used as granules by cutting the bars under petroleum ether. It suffers from the disadvantage that the metal surface rapidly becomes coated with an inert material so it should not be used unless the solvent is pre-dried. Sodium reacts with benzophenone to give a dark blue ketyl radical which is protonated by water to give colourless products. Thus the sodium-benzophenone system is particularly convenient because it is self-indicating, and it is the preferred reagent for rigorous drying of diethyl ether, THF, DME, and other ethereal solvents. Sodium-potassium alloy has been recommended because it is liquid and therefore its surface does not become coated so easily, but this advantage is outweighed by the increased danger resulting from the use of potassium. Sodium residues can be destroyed by slow careful addition of ethanol until hydrogen evolution ceases. The mixture should then be stirred well, to ensure that no coated lumps of sodium remain, before carefully adding methanol. After leaving for several hours the mixture should be stirred again to ensure that all of the sodium has been consumed, and then the mixture should be added cautiously to a large excess of water before disposal.

Sodium should never be added to chlorinated solvents because a vigorous or explosive reaction could occur.

Sodium sulphate, Na_2SO_4

Anhydrous sodium sulphate is a weak drying agent suitable only for drying extracts. It is preferable to magnesium sulphate for drying very acid sensitive compounds.

5.4 Drying of solvents

Solvents may be dried in individual batches using conventional distillation apparatus (Chapter 11), but it is more convenient to dry common solvents such as dichloromethane, diethyl ether, and THF in continuous stills (Section 5.5). In either case the solvent must be protected from moisture using an inert atmosphere (nitrogen or argon). Rigorously dried solvents must be stored under an inert atmosphere and handled using syringe or cannula techniques (Chapter 6).

Acetone

Acetone is completely miscible with water and its susceptibility to acid and base catalysed self condensation makes it particularly difficult to dry. Good results are obtained by drying over 3A sieves (10% w/v) overnight (any longer causes significant condensation), stirring over boric anhydride (5% w/v) for 24h, and then distilling.[1(c)] Distillation after 24h over boric anhydride or 6h over 3A sieves provides material which is adequate for most purposes.

Acetic acid

(Causes burns) Acetic acid is very hygroscopic. It can be dried by adding acetic anhydride (3% w/v) and distilling (b.p. 118°C). Reagent grade acetic acid usually contains some acetaldehyde. If this is likely to cause problems add chromium trioxide (2% w/v) as well as acetic anhydride before distilling, or use analytical grade material.

Acetonitrile

(Toxic) Preliminary drying is accomplished by stirring over potassium carbonate for 24h. A further 24h over 3A sieve or boric anhydride gives moderately dry solvent (~ 50ppm) but much better results are obtained by stirring over phosphorus pentoxide (5% w/v) for 24h and then distilling.[1(a)] Drawbacks of this method are the formation of substantial quantities of coloured residue, and the possibility that the product is contaminated with traces of acidic impurities. If the acetonitrile is required for use with very acid sensitive compounds it is best to redistil it from potassium carbonate.

Ammonia

Distil from the cylinder into a flask cooled to <-40°C and fitted with a dry-ice condenser (Chapter 9). Add pieces of sodium until the dark blue colour persists, and then distil the ammonia into your reaction vessel.

Benzene

(Carcinogenic) Benzene, like most hydrocarbons is very easy to dry. No preliminary drying is required and several reagents will reduce the water content to < 1ppm. Alumina, calcium hydride, and 4A sieves (all 3% w/v for 6h) are the most convenient drying agents and the benzene is then distilled and stored over 4A sieves[1(a)]. Alternatively benzene may be dried over calcium hydride in a continuous still. Toluene may be dried in the same way.

tert-Butanol

Reflux over calcium hydride (5% w/v) and distil onto powdered 3A sieves.[1(g)] Other low molecular weight alcohols, but not methanol, can be dried in this way.

Carbon disulphide

(Highly flammable, toxic)   Distil (using a water bath) from calcium chloride or phosphorus pentoxide (2% w/v). *Do not use sodium or potassium.*

Carbon tetrachloride

(Carcinogenic) See chloroform.

Chlorobenzene

See dichloromethane

Chloroform

(Toxic)   Perhaps the simplest procedure is to pass the chloroform through a column of basic alumina (grade I, 10g per 14ml). This removes traces of water and acid and also removes the ethanol which is present as a stabiliser. Carbon tetrachloride may be purified in the same way. Larger volumes of either solvent can be dried with 4A sieves, or by distillation from phosphorus pentoxide (3% w/v). Distilled chloroform should be stored in the dark to prevent formation of phosgene.

   Addition of sodium to chloroform or carbon tetrachloride may cause an explosion. Chloroform may also react explosively with strong bases and with acetone.

Cyclohexane

See petroleum ether.

Deuteriochloroform (chloroform-D)

This most commonly used solvent for measurement of nmr spectra often contains traces of acid, which can be removed by filtration through a small column of basic alumina (as for chloroform) in a Pasteur pipette.

Decalin (decahydronaphthalene)

Decalin is very easy to dry but it forms peroxides on prolonged contact with air so it is advisable to use a drying agent which will reduce the peroxides.

Reflux over sodium for 2h and distil onto 4A sieves. Tetralin should be treated similarly.

1,2-Dichloroethane

See dichloromethane.

Dichloromethane

Reflux over calcium hydride (5% w/v) and distil onto 4A molecular sieves. Chlorobenzene and 1,2-dichloroethane can be dried in the same way. Dichloromethane can be dried over calcium hydride in a continuous still.

Never add sodium or powerful bases to chlorinated solvents - an explosion may occur. Reaction of azide salts with dichloromethane results in the formation of explosive azides.

Diethyl ether (ether)

(Flammable)   Ether itself, along with the other commonly used ethereal solvents, THF, DME, and dioxan, can contain substantial amounts of peroxides formed by exposure to the air. **These peroxides can cause serious explosions.** Test for the presence of peroxides by adding 1ml of the solvent to 1ml of a 10% solution of sodium iodide in acetic acid. A yellow colour indicates the presence of low concentrations of peroxides whereas a brown colour indicates high concentrations. Low concentrations of peroxides must be removed before further purification and a number of methods have been suggested.[4] Frequently recommended procedures include shaking the solvent with concentrated aqueous ferrous sulphate, or passage through a column of active alumina (which also achieves a degree of drying). Ethers are usually pre-dried over calcium chloride or sodium wire, and rigorously dried over sodium-benzophenone. Careful preliminary drying is necessary because all of these solvents can dissolve substantial quantities of water. The pre-dried solvent is then placed in a reflux apparatus or a continuous still and sodium pieces (1% w/v) and benzophenone (0.2% w/v) are added. The mixture is refluxed under an inert atmosphere until the deep blue colour of the ketyl radical anion persists. The ether may then be collected or distilled onto 4A sieves. This drying method also removes the peroxides which are a serious hazard when handling ethereal solvents. If a continuous still is used it will be necessary

4.   D.R. Burfield, *J. Org. Chem.*, 1982, **47**, 3821.

to add more sodium and benzophenone occasionally. Eventually the still will become very murky as the benzophenone reduction products accumulate. If this happens, or if the blue colour no longer persists, it is time to distil most of the solvent - do not distil to dryness. The sodium residues may then be destroyed as described in Section 5.3. Purified ethers are very susceptible to peroxide formation, so they should be stored in dark bottles under an inert atmosphere, and they should not be kept for more than a few weeks.

1,2-Dimethoxyethane (DME)

See diethyl ether. This solvent is difficult to dry rigorously using sodium-benzophenone and it is recommended that this solvent be distilled *twice* from sodium-benzophenone.

Dimethylformamide (DMF)

Stir over calcium hydride or phosphorus pentoxide (5% w/v, overnight), filter, and distil (56°C at 20mmHg) onto 3A sieves.[1(c)] If phosphorus pentoxide is used a second distillation from potassium carbonate may be necessary. Alternatively, sequentially dry over three batches of 3A sieves (5% w/v, 12h.).

Dimethyl sulphoxide (DMSO)

Distil (75°C at 12mmHg) discarding the first 20% and sequentially dry with two batches of 4A sieves (5% w/v, 12h).[1(c)] Store over 4A sieves.

Dioxan

See diethyl ether.

Ethanol

Distil absolute alcohol from magnesium (as described for methanol) onto 3A molecular sieve powder or sequentially dry over two batches of 3A sieve powder (5% w/v, 12h).[1(g)]

Ether

See diethyl ether.

Ethyl acetate

Distil from potassium carbonate onto 4A sieves.

Hexamethylphosphoric triamide (HMPA)

(Carcinogenic)   HMPA is very difficult to dry and even when stored over 4A sieves it needs to be dried afresh within a couple of weeks. Dry HMPA can be obtained by distilling (89°C at 3mmHg) from calcium hydride (10% w/v) onto 4A molecular sieve (20% w/v).[1(c)]

Hexane

See petroleum ether.

Methanol

(Toxic)   To dry 1 litre of methanol place magnesium turnings (5g) and iodine (0.5g) in a 2 litre flask fitted with a reflux condenser and add methanol (50ml). Warm the mixture until the iodine disappears, and if a stream of bubbles (hydrogen) is not observed add more iodine (0.5g). Continue heating until all of the magnesium has been consumed and then add the remainder of the methanol. Reflux the mixture for 3h, distil (bumping) onto 3A sieve beads (10% w/v), and allow to stand for at least 24h.[1(g)]

Nitromethane

Dry by standing over calcium chloride, filtering, and distilling onto molecular sieves. Do not use phosphorus pentoxide.

Pentane

See petroleum ether.

Petroleum ether (petrol)

(Flammable)   This confusing name is used for mixtures of aliphatic hydrocarbons containing smaller amounts of aromatic compounds. It is generally supplied as several fractions each having a 20°C boiling range (40-60, 60-80 etc.). Alkane mixtures which do not contain aromatic compounds are supplied as pentane, hexane, cyclohexane etc. All of these solvents are readily dried by distilling, and standing over activity grade I alumina (5% w/v), or over 4A molecular sieves.

Pyridines

(Toxic)   Distil from calcium hydride onto 4A molecular sieves.[1(e)]

Tetrahydrofuran (THF)

See diethyl ether.

Tetralin (tetrahydronaphthalene)

See decalin.

Toluene

See benzene.

Xylene

See benzene.

5.5 Solvent stills.

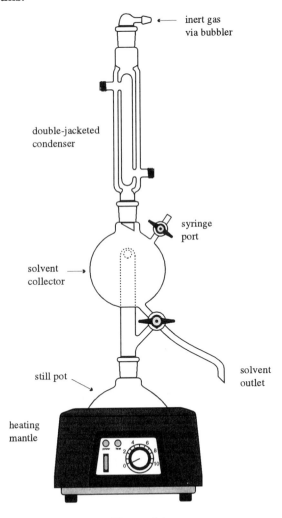

Figure 5.1

There are two main types of solvent still that are commonly employed in organic research laboratories. One is the classical distillation set-up consisting of distillation pot, still-head, thermometer, condenser, receiver-adaptor, and collection vessel. This arrangement is described in more detail in Chapter 11, and is used for the distillation of solvents that are either required infrequently, or that can be stored without deterioration for long periods of time. The other is a continuous still set-up that consists of a distillation pot, collecting head, and condenser (see Fig. 5.1).

This type of still arrangement is used for solvents that are required on a regular basis and the still system is usually left set up, although generally it is only turned on when the solvent is required - *it is not recommended that any type of solvent still is left on unattended for prolonged periods of time* .

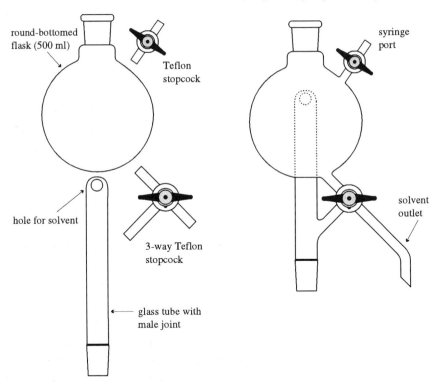

round-bottomed
flask (500 ml)

Teflon
stopcock

syringe
port

hole for solvent

3-way Teflon
stopcock

solvent
outlet

glass tube with
male joint

Figure 5.2

Continuous still systems typically have an upright arrangement that takes up less space than the conventional still set-up and a collecting vessel that is positioned between the still-pot and the condenser. The apparatus is designed such that the distilling solvent is condensed and collects in a collecting head. Once the collecting head is full the solvent simply overflows back into the still pot, allowing continuous distillation without the still boiling dry. The solvent can be drawn off from the collecting head when required, or poured back into the still pot if not.

A typical design for the continuous still collecting head is outlined in Fig. 5.2, and can be simply constructed from a round-bottom flask, ground glass cone, 2-way tap, and 3-way tap. The 2-way tap allows the solvent to be withdrawn *via* syringe which is particularly convenient for anhydrous solvents. The 3-way tap allows the solvent to be collected, drawn off, or poured back into the distillation pot. Obviously the size of the still depends upon the quantity of solvent required, however the still head should always be smaller in capacity than the still pot, so as to avoid the possibility of the still boiling dry.

When setting up a continuous still it is necessary to ensure that the solvent condenser is efficient, usually a double walled condenser is required (see Chapter 9), this is especially true for the lower boiling solvents. Also, all ground glass joints should be fitted with teflon sleeves to ensure a good seal, and prevent jamming. It is inadvisable to use grease on the joints since this will be leached out by the hot solvent, contaminating the solvent and causing the joints to stick. Similarly Teflon taps are to be preferred over glass taps in the collecting head.

If an inert atmosphere is required for the solvent distillation, as is often the case for anhydrous solvents, then the continuous still system requires to be connected to a nitrogen or argon line (see Chapter 9). It is important when using such lines, you ensure that oil bubblers or similar devices do not suck back when the still is cooling. This often happens as a result of the gas volume of the still contracting as it cools, which creates a partial vacuum in the system. It is simply remedied by turning up the flow rate of the inert gas for the period whilst the still is cooling. It is also important that you do not heat up the still system without some mechanism for allowing the increase in gas volume to be released otherwise an explosion may result, this is prevented by incorporating an oil bubbler in the gas line (see Chapter 9).

The design of collecting heads can vary, for instance they can also be constructed using a conical flask instead of the round-bottom flask (Fig.

5.3a). Also, a more complex arrangement (Fig. 5.3b) incorporating a condenser to cool the distillate (particularly useful for high boiling solvents) is often used. This system also has the advantage of less ground glass joints in the set-up, although this can make cleaning the still head more difficult.

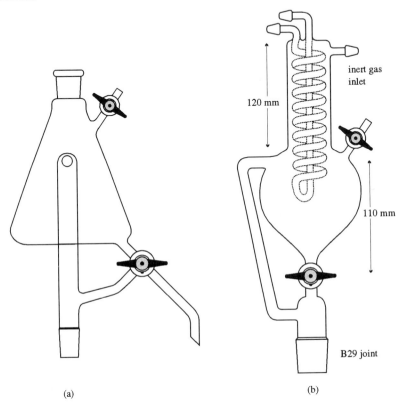

(a)   (b)

Figure 5.3

CHAPTER 6

Reagents: Preparation, Purification and Handling

6.1 Introduction

One of the key steps to becoming a successful practitioner of modern organic chemistry is knowing how to handle and store air- and moisture-sensitive reagents, with the certainty that they have not been contaminated. This is a skill which takes some time to acquire and many people unfortunately learn the hard way, after a string of failed reactions. An efficient, organic chemist achieves good results rapidly, not by cutting corners, but by rigidly observing strict working practices that allow sensitive reagents to be used with confidence. The biggest waste of research time stems from employing reagents or procedures which you *think* are 'OK'. It will usually take far more time to repeat a reaction than it would have taken to re-purify a suspect reagent before starting. Perhaps more important than the time wasted by failed reactions, is the uncertainty which is introduced by using suspect reagents. No matter how carefully the outcome of an experimental reaction is quantified, in terms of yield, stereoselectivity, by-products, etc., the data will be meaningless if there was any uncertainty about the reaction conditions or reagents.

Handling sensitive reagents confidently is not difficult once a few standard techniques are learned and adhered to. In this chapter we will give examples of simple general methods which can be employed to handle a variety of reagents.

6.2 Classification of reagents for handling

The methods used to handle a particular reagent will be dependent on the properties of that reagent and you should be fully conversant with these before you start work. If you are working with a reagent and you are not familiar with its properties, you should look them up, or consult someone who is familiar with them, before you begin. This is very important if you are to use the reagent effectively, without causing a hazard to you and others around you. Once you are conversant with the properties of a particular reagent, the handling requirements are largely a matter of common sense. For example, it is clearly unnecessary to rigorously exclude atmospheric moisture from a reagent which is to be reacted in aqueous solution, but it is important if the reagent is pyrophoric! Most reagents fall into one of the four categories below.

Stable non-toxic reagents

Reagents which are neither sensitive to the atmosphere nor toxic are normally stored in ordinary bottles on the shelf and they are usually easy to handle. However, if you are going to use this type of reagent for reactions with air sensitive materials, *they must also be dry and be stored with the atmosphere excluded!* It is no use carefully setting up an anhydrous reaction, then adding a reagent straight from an unsealed bottle on the shelf.

Stable reagents which are toxic or have an unpleasant odour

These should be treated in the same way as those in the previous category, except that special precautions should be taken to avoid their escape into the lab. They should always be used in a fume cupboard and stored in a ventilated storage cupboard.

Reagents which will decompose on exposure to moisture or air

These reagents should always be stored in special containers, under a dry inert atmosphere. Whenever they are used, they should be measured and transferred using techniques which constantly maintain the inert atmosphere. Some of these techniques will be described below.

Reagents which decompose explosively or pyrophorically on exposure to moisture or air

These should be treated as above, but extra care should be taken, especially when dealing with residues once the reaction is over.

6.3 Techniques for obtaining pure and dry reagents

Before using any organic reagent or starting material for a reaction its purity should be checked. It should not be taken for granted that a reagent is pure simply because it has been obtained from a commercial source. Indeed the specification of many commercial compounds is much less than 100% and this should always be checked. Even if the specified purity of the reagent is high, at least one analytical check should also be carried out, such as tlc, gc or nmr. This check is particularly important if you are attempting new reactions. If the purity of the reagent is not up to the level required, standard purification techniques (distillation, recrystallization, chromatography, sublimation etc.) should be used to purify it (see Chapter 11 for more details).

Remember that any reagent which is to be used in a reaction under inert conditions with sensitive reagents, must also be dried very carefully!

Once you have gone to the trouble of purifying and drying a reagent, it is good practice to store it very carefully in order to keep it dry for future use. This normally means storing it in a sealed container, *under an inert atmosphere.* When you keep reagents in this way, you *must* be able to rely on their purity and it is therefore best to keep them for your own personal use. Remember that a person who has not gone to the trouble of purifying a reagent is not likely to take good care of yours!

In this section we will describe some typical techniques for purification and drying of reagents. There are several other more specialized texts which should be consulted if you need to purify other reagents.[1,2,3]

6.3.1 Purification and drying of liquids

For most liquids the method of choice for both purification and drying is distillation. In many cases the liquid is either dried over a drying agent before distillation or distilled from a drying agent (see also Chapter 5).

1. *Purification of Laboratory Chemicals*, 3rd ed., D.D. Perrin and W.L.F. Armarego, Pergamon Press, Oxford, 1988.
2. *Reagents for Organic Synthesis,* Vols. 1-13, Fieser and Fieser, Wiley, New York.
3. *Organic Synthesis*, Col. Vols. 1-6, Wiley, New York.

Distillation under inert atmosphere at normal pressure

This technique is used to purify and/or dry most liquid reagents which boil at less than about 150°C, at atmospheric pressure. Typically the liquid is pre-dried by shaking over magnesium sulphate, then decanted onto a more active drying agent, such as powdered calcium hydride, from which it is distilled. It is important to make sure that the drying agent does not react with the liquid. If you require a dry reagent, it is useless to carry out the distillation in the atmosphere, and the process must be carried out under argon (or nitrogen). For ease and reliability we suggest that a one-piece distillation apparatus is used for this type of distillation, in conjunction with a double manifold; the technique is fully described in Chapter 11.

When the distillation is complete remove the collector, seal it quickly with a septum and purge the bottle with inert gas. For compounds which react with rubber, such as Lewis acids, a Teflon stopper should be used. Some of the less sensitive reagents can simply be poured into a reagent bottle before it is sealed, provided this is done quickly. However, if the reagent is particularly sensitive to air or moisture (such as a Lewis acid), a cannulation technique should be used to transfer it. See Section 6.4.1 for more details about storage and transfer of reagents under dry, inert conditions. Table 6.1 gives some examples of common reagents which should be distilled under an inert atmosphere.

Table 6.1

Reagent	Drying agent	b.p.°C	Comment
Methyl iodide	$CaCl_2$ before dist.	43	Very toxic! Wash with $Na_2S_2O_3$ or pour through alumina first to remove I_2
Diisopropylamine	dist. from CaH_2	84	
Triethylamine	dist. from CaH_2	89	
Isobutyraldehyde	$CaSO_4$ before dist.	62	readily oxidized
Titanium tetrachloride	none	136	very reactive with moisture
Cyclohexene	dist. from Na	83	wash first with $NaHSO_3$ to remove peroxides
Crotonaldehyde	$CaSO_4$ before dist.	104	use vigreux

Distillation under reduced pressure

For liquids with a high boiling point, or which decompose on heating, distillation under reduced pressure is the usual method of purification. Again it is convenient and effective to use a one-piece distillation apparatus in conjunction with a double manifold for this type of distillation. The procedure is described in more detail in Chapter 11, and some examples of reagents which can be distilled under reduced pressure are given in Table 6.2.

When the distillation is complete the heat is turned off and the two-way tap on the double manifold is slowly turned to the inert gas position. The flask containing the distillate will then be conveniently under an inert atmosphere and should be quickly removed and fitted with a tightly fitting septum (see Section 6.4.1 for more details about storing and transferring reagents under inert conditions).

Table 6.2

Reagent	Drying agent	b.p.°C/mm	Comment
Dimethyl formamide	$MgSO_4$ before dist.	76°/39	do not use CaH_2 or other basic drying agents
$BF_3.Et_2O$	dist. from CaH_2	67°/43	reacts with moisture
$SnCl_4$	dist. from tin	high vac.	reacts with moisture
Et_2AlCl	dist. from NaCl	107°/25	heated nichrome spiral in 50cm column; reacts with air - spontaneously flammable!
Et_3Al	none	130°/55	as above
Benzaldehyde	$MgSO_4$ before dist.	62°/10	wash with Na_2CO_3 before dist. Store over 0.1% hydroquinone
Benzyl bromide	$MgSO_4$ before dist.	85°/12	dist. in dark; *lachrymator*

Some special cases for drying of liquids

There are a number of liquid reagents which hydrolyse when they come in contact with water and subsequently contain the hydrolysis product, which is usually an acid. These reagents are distilled from a high-boiling amine, or other base, so that any acidic impurity is removed. It is always important that the distillation is carried out under inert atmosphere (or under reduced pressure) and the techniques are exactly as above. Some examples are given in Table 6.3.

Table 6.3

Reagent	Distil from	b.p.°C	Comment
Acetic anhydride	quinoline	138	10:1 with quinoline
Acetyl chloride	quinoline	52	10:1 with quinoline
Me$_3$SiCl*	quinoline	106	10:1 with quinoline
(Chlorotrimethylsilane)			

*Another method for removing HCl from trimethylsilyl chloride (chlorotrimethyl silane) is to add it to an equal volume of triethylamine in a centrifuge tube, sealed under argon with a septum. After centrifuging, triethylamine hydrochloride forms a solid precipitate and the liquid can be drawn off by syringe and used for most purposes as 50% Me$_3$SiCl.

Details about how to store and handle liquid reagents under dry and inert conditions are given in Section 6.4.1.

6.3.2 Purifying and drying solid reagents

If the purity of the reagent is not up to the level required, standard solid purification techniques (chromatography, recrystallization, sublimation) should be used to purify it (see Chapter 11 for more details).

Remember that if a solid reagent is to be used in a reaction under inert conditions with sensitive reagents, it must be dried very carefully. For bulk drying of solids an oven is often used (preferably a vacuum oven), or the solid is placed in a vacuum desiccator over a drying agent (P$_2$O$_5$, H$_2$SO$_4$, etc.). If the reagent is to be stored, for future use in reactions under inert conditions, it should be kept under argon (or nitrogen), in a sealed bottle, to prevent exposure to moisture. Small quantities of solid can often be dried in the reaction flask, prior to the apparatus being set up. This is conveniently done by connecting the dried flask, containing the solid, to the double manifold, evacuating under high vacuum for several hours, then introducing argon (or nitrogen).

Purification of some common solid reagents

AIBN (α,α'-Azobis(isobutyronitrile))

Recrystallize from diethyl ether and dry under vacuum, over P_2O_5 at room temperature. Store under inert atmosphere, in the dark, at -10°C.

Para-toluenesulphonyl chloride

Para-toluenesulphonyl chloride often contains a considerable quantity of *para*-toluenesulphonic acid. This can be removed by placing the reagent in the thimble of a soxhlet apparatus containing dry petroleum ether. After several hours of extraction under an inert atmosphere, the chloride will have dissolved in the solvent and the unwanted acid will be left behind in the soxhlet thimble. On cooling the solvent mixture, the acid chloride crystallizes and can be collected by filtration. The purified material should be stored under an inert atmosphere.

Copper(I) iodide (for the preparation of cuprate reagents)

Impure copper iodide is often slightly brown because of contamination by iodine and copper(II) salts. It is very important that these contaminants are removed and that the reagent is dried efficiently if it is to be used in the preparation of copper-lithium reagents. This is accomplished by placing the material in the thimble of a soxhlet apparatus and extracting with methylene chloride for several hours (usually overnight) until no further colour is being removed. The almost white reagent is then dried under vacuum and stored under argon (or nitrogen) in a dark bottle, sealed with parafilm.

Magnesium (for Grignard reactions etc.)

Wash with ether, to remove grease from the surface of the metal, dry at 100°C under vacuum and cool in a desiccator. For a more active form of magnesium, stir the turnings under nitrogen overnight. They will turn almost black as the oxide coat is removed from the surface. The material produced is very active and should be used immediately or stored for a short period under inert atmosphere.

Zinc

Zinc is normally coated with oxide, which must be removed prior to use. This can be done by stirring with 10% HCl for 2 min, then filtering and washing with water, followed by acetone. The metal can then be vacuum-dried and kept under an inert atmosphere prior to use.

6.4 Techniques for handling and measuring reagents

Although extreme care must be taken to exclude air and moisture from reagents kept under an inert atmosphere, once you become familiar with the simple techniques outlined in this section, you should find them easy to use. Bottles and flasks containing liquids or solutions sealed under an inert atmosphere are widely used in modern organic chemistry. Examples of such reagents are: organolithium reagents; lithium aluminium hydride and other active hydride reducing agents; Lewis acids; etc. A knowledge of the techniques required to manipulate such reagents, without allowing them to come into contact with the atmosphere, is therefore essential.

When handling air sensitive reagents always *think ahead, and design the whole sequence of events you intend to follow so that air never comes into contact with the reagent.*

Methods for preparation and titration of alkyllithium reagents is given in Section 6.5 of this chapter and a method for titration of hydride reagents is given in Chapter 7, Section 7.4.

6.4.1 Storing liquid reagents or solvents under an inert atmosphere

After drying and/or distillation, reagents or solvents can be stored, under argon, in a bottle or flask sealed with a septum. There are also many other commonly used reagents that are extremely reactive towards water and/or oxygen and these are stored in the same way. Examples of such reagents are: alkyllithium reagents; Grignard reagents; organoboranes; metal hydrides; organoaluminium compounds and Lewis acids. Some of these are available commercially and are delivered under inert atmosphere in sealed bottles. Less common reagents may be prepared in the lab and then stored for future use. The seals of the commercially supplied containers have a limited life-span and if one is suspect it should be replaced with a rubber septum. For Lewis acids, which often react with rubber, a Teflon stopper can be used. For storing small quantities of liquid under inert atmosphere 'Mininert valves' are very useful. These screw threaded bottle caps incorporate Teflon valves through which a syringe needle can enter the container. They provide a better seal than ordinary septa (see Section 6.4.4 for more detail on the use of Mininert valves).

Preparing a storage vessel or reaction fitted with a septum, under inert gas

1. Choose a septum which fits tightly into the neck of the flask or bottle that you wish to seal.
2. Dry the container thoroughly, either in an oven or by heating under vacuum.
3. Fit a new septum as quickly as possible, minimizing exposure of the reagent to the atmosphere.
4. Insert a needle through the septum to act as a vent.
5. Quickly inject inert gas into the flask using a needle connected to an inert gas line by a Luer adapter (Fig. 6.1a).
6. Once the vessel has been thoroughly purged with inert gas, remove the vent, then the inert gas needle.

For long-term storage, the septum should be secured with copper wire and for extra protection another septum of the same size, without any needle holes in it, can be pulled over the first (Fig. 6.1b and c). The second septum is removed before syringing liquid from the vessel. It is also a good idea to wrap parafilm around the septum, especially if the bottle is to be stored in the refrigerator. Moisture sensitive reagents can be added to the vessel by cannulation as described below.

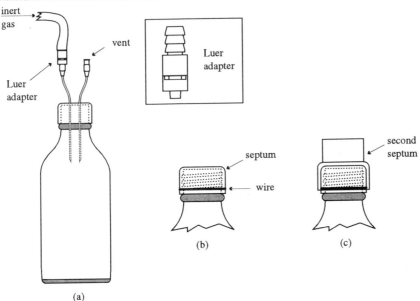

Figure 6.1

Changing a septum or fitting a septum to a bottle containing a liquid

Non-reactive reagents or solvents may be poured into a bottle that has been dried, then fitted with a septum and purged with inert gas as described in steps 3-6 above. An existing septum can be replaced in similar manner, provided the new septum is fitted quickly, minimizing exposure of the reagent to the atmosphere.

A reagent or dry solvent stored under inert atmosphere, as described above, and used carefully, according to the techniques below, can be kept for many months and used many times over. If a reagent bottle has been stored in a refrigerator or freezer, always allow it to warm up to room temperature and remove any condensation, before unsealing it.

6.4.2 Bulk transfer of a liquid under inert atmosphere (cannulation)

As a general rule, never attempt to remove liquid from a container which is sealed under inert gas, unless you have pressurized the container with inert gas first (Fig. 6.2a). *Also, whenever you are using ground glass joints connected under pressure always secure them with either plastic (Bibby-type) clips, elastic bands or springs.*

General principle of cannulation and gas pressure adjustments

Cannulation is a general procedure for the transfer of a liquid from one container to another whilst maintaining an inert atmosphere. The principle of cannulation is very simple. A positive pressure is applied to the flask from which the liquid is to be transferred. This pressure forces the liquid out through a double ended needle (cannula) into the receiving flask (see Fig. 6.3). In order for the liquid to flow, there must be some means by which the gas in the receiving vessel can escape. The simplest way to allow this to happen is to vent the receiving flask with a short needle passed through the septum. It is good practice to connect the vent needle to a bubbler to prevent suck back of air occurring.

It is quite common that the flask into which you need to cannulate the liquid is already part of an inert gas system and can not simply be vented. In this case the pressure applied to the first flask must be higher than that in the receiving flask and *there must be some means by which gas can escape from the receiving flask.* The inert gas system to which the receiving flask is connected should incorporate a bubbler (for example, see Fig. 3.7) and this automatically provides a means for escape of gas. To create a higher pressure in the delivery flask and control the rate of flow of liquid during cannulation, the vent of the bubbler in the delivery system is restricted with a finger. For more precise control of pressure the vent can be restricted by connecting a Rotaflow tap or needle valve to it (see Fig. 6.2).

General procedure for cannulation

The following procedure is a general method for transferring liquids by cannulation and will be referred to in other sections of the book.

1. Make sure the bottle or flask into which you are going to transfer the reagent is thoroughly dry, fit a septum into the neck and purge with inert gas as described in the previous section. If the container was oven dried it is preferable to fit the septum whilst it is still hot and allow it to cool as it is being purged with inert gas (Fig. 6.1a).

2. When the bottle or flask is cool, make sure the septum is fitted tightly and wire it on, then remove the vent *before removing* the argon line (Fig.6.1b). The container now contains an inert atmosphere and is ready for use.

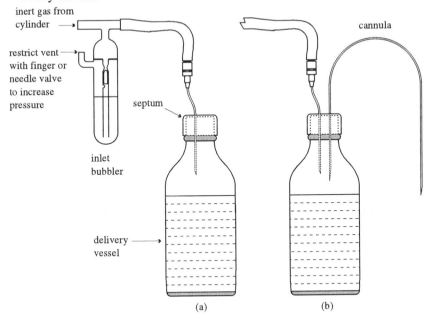

Figure 6.2

3. Insert an inert gas line needle into the septum of the bottle or flask containing the liquid to be transferred. If you are transferring liquid to a flask that is already part of an inert gas system, a separate gas line must be used to pressurize the bottle and a simple bubbler system is recommended (Fig. 6.2a).

4. Insert a double ended needle (cannula) into the bottle containing the reagent, *taking care not to push the needle below the surface of the liquid at this time.* Check that the inert gas is flowing through the system and out through the cannula (Fig. 6.2b).

5. Insert the other end of the cannula through the septum of the new bottle or flask, then push the inlet end below the surface of the liquid in the delivery flask. There should be no flow initially, because the receiving container is sealed. Vent it by inserting a short needle through the septum. The vent needle can be connected to a bubbler or an inert gas system which incorporates a bubbler. If the liquid does not start to flow, increase the pressure by restricting the vent of the inlet bubbler using a finger, a septum or by connecting a needle valve (Fig. 6.3). When argon is used it is normally safe to dispense with the bubbler connected to the vent of the receiving flask.

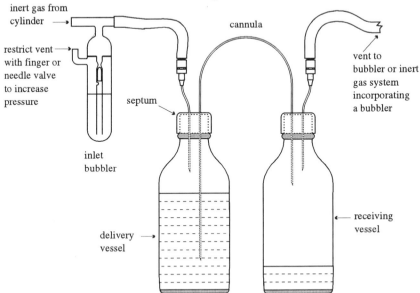

Figure 6.3

6. When all the liquid has been transferred, first remove the vent needle, then remove the cannula at the receiver end. The remainder of the system can then be dismantled in any order, but remember to be very careful with the residues, particularly if the liquid is toxic or reactive to moisture. Take extreme care when transferring very reactive reagents, such as ^tbutyllithium, which is pyrophoric on contact with moisture.

6.4.3   *Using cannulation techniques to transfer measured volumes of liquid under inert atmosphere*

The cannulation technique can easily be adapted for measuring volumes of liquid under inert atmosphere. It is mainly used for measuring quantities of liquid that are too large to be conveniently handled by syringe (> 50ml), but it is also suitable for measuring pre-cooled solutions without significant warming. The technique is identical to that described in the previous section, but an intermediary graduated container is used. The graduated container can be a measuring cylinder with a neck that can be fitted with a septum (Fig. 6.4a), or it can be a graduated Schlenk tube (Fig. 6.4b).

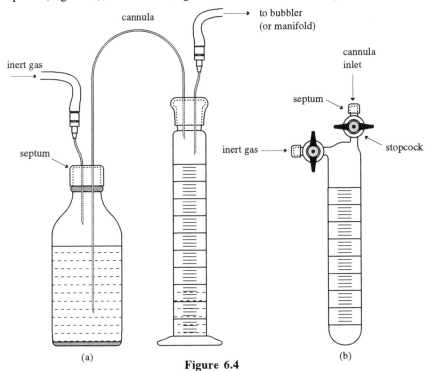

Figure 6.4

1. Dry the graduated container and transfer the required quantity of liquid
    to it according to the procedure described in Section 6.4.2 (Fig 6.4a).
2. Once the required volume of liquid has been measured, remove the
    needle and bubbler from the storage bottle and insert it through the
    septum of a dry receiving vessel that has been filled with inert gas
    (Section 6.4.1).
3. Apply inert gas pressure to the graduated cylinder and vent the
    receiving flask to deliver the liquid (Fig. 6.5). See also Chapter 9 for
    more details on setting up reactions under inert conditions).

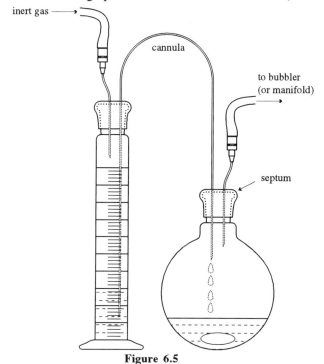

Figure 6.5

If the liquid is being measured for addition to a reaction flask, an
alternative procedure is to use a graduated pressure equalizing dropping
funnel attached to the apparatus. The required quantity of liquid can then be
cannulated into the dropping funnel directly.

Types of cannula

For most purposes cannulation can be carried out using an ordinary double
ended needle, bent to a suitable shape (Fig. 6.6a). A cannula made by
joining two long syringe needles to a Luer to Luer stopcock allows the flow

of liquid to be controlled (Fig. 6.6b). For transferring large volumes of liquid the 'flex-needle' (available from Aldrich Chemical Co.) is useful. This is a jacketed Teflon tube with a wide-bore needle attached to either end (Fig. 6.6c). It provides rapid liquid transfer, and a degree of insulation for a cold reagent. The wide-bore needles of the flex-needle should not be inserted through a new septum without first making a hole with a normal gauge needle.

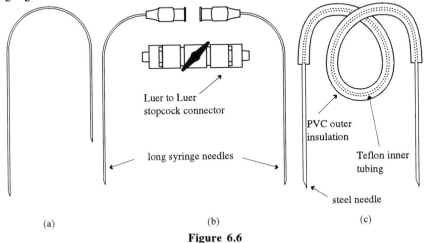

Luer to Luer
stopcock connector

long syringe needles

PVC outer
insulation

Teflon inner
tubing

steel needle

(a)   (b)   (c)

Figure 6.6

For small scale work all Teflon cannulas are very useful, and some people prefer them to syringes. They are easily made by taking a piece of narrow-bore (1.5mm) Teflon tubing, of a suitable length and cutting each end at a shallow angle with a sharp razor blade or scalpel (Fig. 6.7). Always be sure to make holes in septa, using a metal needle, before attempting to insert a Teflon cannula.

ends cut at an angle
using a scalpel

narrow bore
Teflon tubing

Figure 6.7

6.4.4 Use of syringes for the transfer of reagents or solvents

Syringes are extremely useful for transferring small quantities (up to 50ml) of air sensitive reagents or dry solvents from one bottle or flask to another. There is a variety of syringe types and you need to appreciate which type is suitable for a particular application. It is most important to choose a syringe of an appropriate size for the quantity of liquid you need to transport. Using a syringe with a volume much larger than you need will lead to inaccuracy and using a syringe which is too small will waste time. For most purposes the syringe should be equipped with a long needle (10-20cm). This is necessary in order to avoid any need to tip the reagent or solvent bottle when syringing from it. When using syringes, the condition of septa should be checked regularly. Their life can be extended by making sure that the needle tips are kept sharp, using emery paper.

The skill of using syringes is of great importance to an organic chemist, but, as with all skills, you will only become proficient after a certain amount of practice. It is a good idea to have your own set of syringes which are appropriate for the type of work in which you are engaged. They will be some of your most frequently used tools so you should practice using them and take great care of them.

Types of syringe

Micro syringes

Micro syringes are extremely useful for small scale synthetic work, delivering small quantities of liquid accurately. Sizes ranging from 10μl to 1ml are common, but 100μl is perhaps the most useful size to have.

There are two general types of micro syringe, gas-tight and liquid-tight. The liquid-tight syringes have matched glass bores and stainless steel plungers, which are not interchangeable (Fig. 6.8).

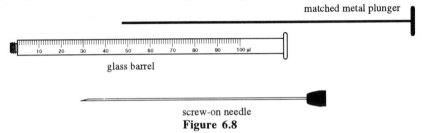

Figure 6.8

Certain types of reagent will corrode the steel plunger and cause seizure if the syringe is not cleaned immediately after use. These are indeed delicate and quite expensive items which should always be treated with great care (see later for care of syringes).

Gas-tight micro syringes have Teflon tipped plungers, making them more inert to chemical reagents (e.g. Lewis acids). The barrels and plungers are also interchangeable and they are thus much more reliable than the liquid-tight syringes. The drawback with gas-tight syringes is that they are more expensive (Fig. 6.9).

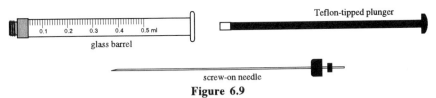

Figure 6.9

Both types of micro syringe are available with either fixed or removable needles, but the type with removable needles are generally preferred. Removable needles have the advantages that they can be replaced when damaged or blocked and that different sizes can be fitted. Long needles for micro syringes normally have to be ordered specially.

All-glass Luer fitting syringes

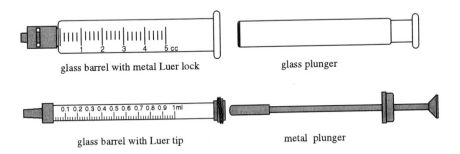

Figure 6.10

All-glass Luer syringes range in size from 1ml to 100ml, are generally quite cheap and are the most commonly used type. Some glass syringes have matched barrel and plunger and they can only be used as a pair, but most modern syringes have interchangeable barrels and plungers. Syringes with interchangeable parts will be marked as such and these are the preferred type to use. Rocket-type syringes have metal plungers and are somewhat more reliable than the all-glass type. Both types are adequate for most purposes, but should be cleaned immediately after use and treated carefully to avoid jamming.

Glass gas-tight Luer syringes

For most purposes these are the best type of syringe available. The barrel of the syringe is made from glass and the plunger is made from metal or plastic and has a separate Teflon tip (Fig. 6.11). Their main advantages over all-glass syringes are that they are less prone to leakage and jamming. They are very expensive, but all the components are replaceable, so they do last a long time if they are looked after.

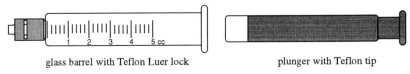

glass barrel with Teflon Luer lock   plunger with Teflon tip

Figure 6.11

Plastic disposable syringes

Plastic syringes are cheap, hold a very good seal, they are quite accurate and virtually unbreakable. They can be used many times over, but they do have the drawback of being susceptible to attack from certain solvents. For this reason they are most widely used for transferring aqueous solutions, such as aqueous hydrogen peroxide.

Syringe fittings

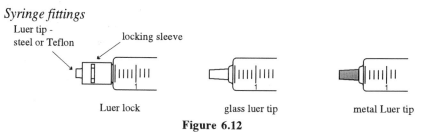

Luer tip -
steel or Teflon   locking sleeve

Luer lock   glass luer tip   metal Luer tip

Figure 6.12

Micro syringes normally have one of two types of screw-on needle, as shown in Figs. 6.8 and 6.9. The fittings of larger syringes are more standard and are normally the Luer type. A Luer fitting is a joint produced with a standard taper which matches the internal taper of a Luer syringe needle. There are several types of Luer fitting (Fig. 6.12).

The simplest fitting is the Luer glass tip, but these break very easily and should be avoided. Metal Luer tips are more robust, but the needles tend to work loose and for this reason Luer-lock syringes are preferred. These have a slot into which the rim of the needle engages so that it can not come loose. The most common fitting is the metal tip Luer-lock, but the best seal of all is provided by the Teflon tip Luer-lock, normally found only on gas-tight syringes. In this case the Luer tip is Teflon and the locking sleeve is metal.

Syringe needles

Luer fitting syringe needles are available in a range of lengths and bore diameters. The optimum bore diameter depends on the volume of liquid being dispensed. As a rough guide we use 22 gauge needles for volumes of less than 1ml, 20 gauge for volumes between 1 to 5ml and 18 gauge for larger quantities. Long needles are usually preferred for transferring liquids by syringe, but short needles are useful as nitrogen inlets or vents. Disposable needles with plastic shanks can also be used for this purpose.

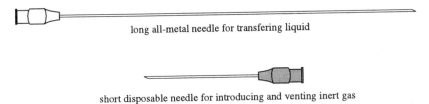

long all-metal needle for transfering liquid

short disposable needle for introducing and venting inert gas

Figure 6.13

Most needles have a 12° bevel and this should be kept in good, sharp condition so that minimal damage is caused when penetrating a septum. Needles can be sharpened by rubbing with fine emery paper. Flat ended needles are sometimes useful for removing the last traces of liquid from a container, but they should only be introduced through septa which have previously been pierced by a sharp needle.

Cleaning and care of syringes

Syringes and needles can give good service for long periods, but careful treatment is required to prevent leakage and jamming. It is therefore very important to clean syringes and needles *immediately* after use. Organic liquids or solutions are generally quite easily removed, simply by flushing with solvent. However, more care is required when a reactive reagent such as butyllithium or titanium tetrachloride has been used. To clear such reactive reagents the syringe should *immediately* be flushed several times with the solvent in which the reagent is dissolved. Whatever the syringe has been used for it should *always* be dismantled for final cleaning of the individual components. If the syringe has been used for an alkyllithium or any other strongly basic reagent, it should be washed with dilute HCl, water, then acetone. After using an acidic reagent or a Lewis acid, the components should be washed with dilute sodium hydroxide, water, then acetone.

If a syringe has jammed or become contaminated with remnants of a stubborn reagent, handle it carefully and *do not use force* to free the plunger. This will almost certainly lead to the syringe being broken and you may well end up with a cut hand also. One of the most reliable methods for removing a contaminant from a syringe, or syringe needle, is to submerge it in a solvent which will dissolve the contaminant and place it in an ultra sonic cleaning bath for a few minutes. Sonication will often free syringes which are completely jammed provided you choose the correct solvent. If you do not know what is blocking the syringe try sequentially sonicating in the following series of solvents: 10% HCl, then water, then dilute NaOH, then water, then acetone, then methanol and finally dichloromethane.

How to handle a syringe

The type of reagent that you will need to deliver by syringe is very often one that has to be measured precisely in order for the reaction to be successful. It is also likely to be a very corrosive reagent, such as butyllithium, which you do not want to splash around the lab or spill onto your skin. So, before attempting to use a syringe for transferring any reagent, practise using it first by measuring solvent until you are confident that you can handle it safely. You will often need to hold a syringe full of liquid in one hand whilst performing another operation with the other hand. One way to do this is as follows:

1.   Fill the syringe with the required volume of liquid using both hands.
2.   Place the syringe against the palm of one hand using the 3rd, 4th and 5th fingers of the same hand to grip the barrel.

3.   Carefully place the forefinger around the plunger and against the end of
      the barrel to hold the plunger firmly in place.
4.   The liquid can now be delivered by pressing the plunger down using
      the thumb, and the other hand is completely free.

This method should be practised using solvent until your technique is good
enough confidently to transfer liquid without spillage.

Preparing a syringe for use

An appropriate clean syringe and needle should be chosen, dried in an oven,
then left to cool in a desiccator. Before using a syringe it should be purged
with argon (or nitrogen) and this is conveniently done by 'syringing out' the
gas *via* a septum inlet attached to the inert gas system. This should be
repeated several times (Fig. 6.14). After purging with argon (or nitrogen)
the syringe can be kept for a short time before use if the tip of the needle is
inserted into a rubber stopper.

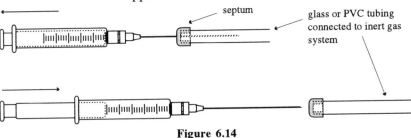

Figure 6.14

Syringing liquid from a bottle or flask under inert atmosphere

Before attempting to syringe liquid from a bottle which is under an inert
atmosphere, make sure you have: an effective inert gas system with a needle
outlet; a dry syringe (and preferably a spare) prepared as described above;
and a receptacle, *also under inert conditions*, into which you are going to
transfer the liquid. The procedure outlined below should be practised with
solvent first if you have not done it before.

1.   Pressurize the reagent container with an inert gas line using a syringe
      needle attached *via* a tubing connector.
2.   Carefully insert the needle of the syringe through the septum, holding
      the syringe plunger in. Then *very slowly* fill the syringe by pulling the
      plunger up (Fig. 6.15a). You should never fill the syringe to more than
      about two-thirds capacity.

3. Tip the syringe upside down, bending the long needle. Slowly force any excess reagent and bubbles back into the bottle until the exact volume is indicated (Fig. 6.15b).

4. Holding the syringe barrel carefully in place with one hand, slowly withdraw the needle from the septum using your other hand to control it. Then, avoiding prolonged exposure to the atmosphere, *quickly* insert the needle into the receptor.

5. Slowly deliver the measured volume into the receiver flask, which should itself be connected to an inert gas system.

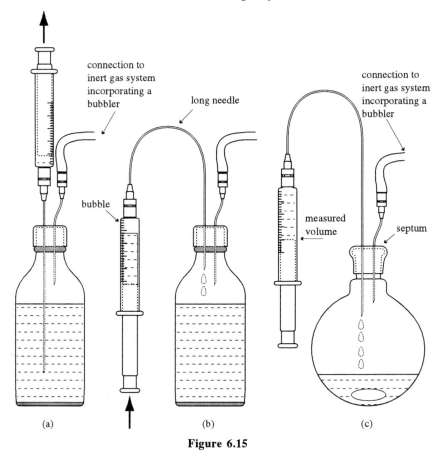

Figure 6.15

Syringing reagents which are extremely moisture sensitive

When a particularly moisture sensitive reagent is being transported by syringe, simply exposing the tip of the syringe needle to the atmosphere may cause problems. For example ᵗbutyllithium can be pyrophoric on contact with atmospheric moisture. However, there is a very simple method for keeping the syringe tip under inert atmosphere at all times. Take a short piece of dry glass tubing (about 5cm long), insert a tight fitting septum in each end and purge with inert gas. This provides a mobile inert gas capsule that can be used to protect your syringe tip (Fig. 6.16a). The procedure for transferring the liquid is exactly as described above except for the following points:

1. When drawing the liquid out of the reagent container the syringe needle is passed through both septa of the inert gas capsule (Fig. 6.16b).
2. After filling the syringe the end of the needle is pulled into the capsule so that the tip is protected (Fig. 6.16c).
3. The capsule is pressed against the septum of the receiving flask and the needle is pushed through both septa to deliver the liquid (Fig. 6.16d).

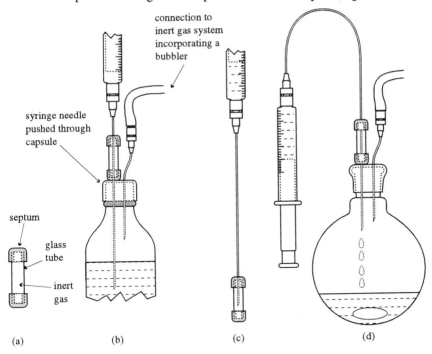

Figure 6.16

Syringing from Mininert bottles

Mininert valves are too narrow for two syringe needles to pass through them. It is therefore impossible for a container fitted with a Mininert valve to be pressurized from an inert gas system while syringing from it. However, because the containers used with these valves are very small a very simple alternative procedure can be used.

1.   After flushing a syringe with inert gas fill it with slightly more gas than the quantity of liquid you wish to syringe. (A gas-tight syringe is preferred for this procedure).

2.   Open the Mininert valve, push the syringe needle through and push the plunger to pressurize the container with inert gas.

3.   With the needle below the surface of the liquid allow the syringe to fill with liquid, under the pressure which you have introduced (never suck liquid into the syringe as this will draw air into the system).

4.   Bend the needle to invert the syringe, displace air bubbles and measure the required volume of liquid.

5.   Remove the needle from the container and deliver the liquid as usual.

6.4.5 Handling and weighing solids under inert atmosphere

Manipulation of solid reagents under dry conditions tends to be somewhat more awkward than handling liquids or solutions. However, there are two distinctly different problems which can be considered. Firstly, there is the use of dry but unreactive solid reagents in reactions under inert conditions, and this will be dealt with in Chapter 9. The other, and more significant problem, is manipulation of very reactive solid reagents which may react with moisture in the atmosphere. The only way to handle solids under completely dry conditions is to use a glove box, which is a sophisticated and relatively expensive piece of equipment. Fortunately, there are very few reagents used routinely in organic synthesis that need to be handled in absolutely dryness. In cases where a particularly reactive solid is to be handled or when great accuracy is required, a simple glove bag can be used. This is essentially a large, clear polythene bag, which can be sealed after equipment and reagents have been placed in it. Inlets are provided for electricity cables, gas lines etc. and glove inserts allow for the manipulation of the apparatus inside the bag. After loading all the equipment and the reagent into the bag, it is evacuated three times and filled with an inert gas (a double manifold can be used for this purpose). Some skill is required to work within a glove bag so, as usual when working under inert conditions, plan your work very carefully making sure everything you require is on hand before you start. It is also a good idea to practice working in the bag before you attempt the 'real thing'.

For handling the vast majority of solid reagents used in organic synthesis a glove box/bag is unnecessary, provided you master a few simple techniques, and work careful and quickly. The most common moisture sensitive solids are metals (sodium, lithium, potassium etc.) and metal hydrides (sodium hydride, potassium hydride, lithium aluminium hydride etc.) and the most commonly encountered problem is weighing. We have already seen that measuring liquids under inert atmosphere is relatively simple and for this reason many air-sensitive organic solids are sold in solution. In order to weigh a solid without using a glove box, a certain amount of exposure to the atmosphere is inevitable, but this can be minimized, provided you are careful. When carrying out a reaction it is normally best if you can plan events so that any air sensitive solid is added to the reaction flask before other reagents or solvent (see Chapter 9 for more details).

Weighing solid reactive metals

As usual, plan your work ahead and make sure you have a flask pre-dried and under inert gas. The metal will normally be under paraffin oil and may also have an oxide or hydroxide layer coating it. Both these impurities will have to be removed before weighing the metal and this can be done using the following sequence (Fig. 6.17). ***Remember that some reactive metals are pyrophoric when they come into contact with moisture.***

1. Place the metal into a beaker, covering it with oil, then cut some of it into small pieces with a scalpel, removing any coating and leaving the shiny surface exposed. ***Heavily coated potassium has been known to detonate on cutting and should be discarded.***

2. Using a pair of forceps, quickly wash the oil from the metal chunks in a second beaker containing *dry* hexane or pentane. Remove the chunks, allowing the solvent to evaporate very briefly, and drop into a weighed beaker of oil.

3. Once you have weighed the required quantity, remove the metal chunks, wash again in the petroleum ether, quickly add them to the reaction flask and re-connect to the inert gas system.

4. When weighing is complete add an alcohol (e.g. ethanol) to the beakers to neutralize any scraps of metal before washing with water.

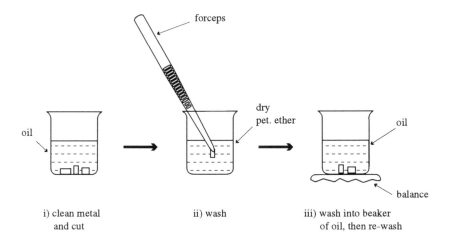

Figure 6.17

Handling metal and metal hydride dispersions

Finely divided metals and metal hydrides are very useful reagents and are usually packed as dispersions in paraffin oil. The most common of these reagents are sodium hydride, potassium hydride and lithium metal. Whilst dispersed in the oil the reagents are moderately stable and can be weighed out quickly in the atmosphere.

For some experiments the dispersion can be used without separating out the oil. To do this simply weigh it into a pre-dried reaction flask and place it under inert atmosphere by sequentially evacuating, then purging the flask with an inert gas. A 3-way Quickfit tap connected to a double manifold is ideal for this purpose.

There are various techniques for removing the oil from a metal dispersion. The simplest method is as follows:

1. Weigh the dispersion in a flask (do not forget to take the oil into account) and place it under inert atmosphere. Use a double manifold connected by a 3-way Quickfit tap if possible (Fig. 6.18a).

2. While maintaining a rapid flow of inert gas through the bubbler of the inert gas system, open the 3-way tap and add some dry petroleum ether using a syringe. Swirl the flask to dissolve the oil, then let it stand until the metal has settled at the bottom (Fig. 6.18b).

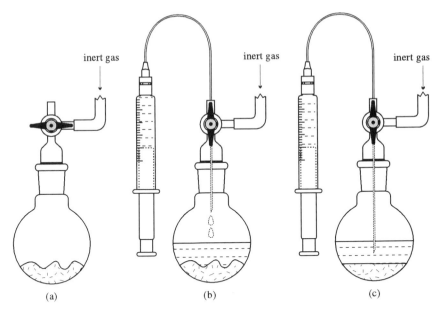

Figure 6.18

3. Draw off the petroleum ether carefully using a syringe (Fig. 6.18c). Discard the solvent carefully, into alcohol, as it may contain a small quantity of the metal.

4. Repeat the washing process two more times.

5. The flask can then be evacuated to remove last traces of petroleum ether, then re-weighed to determine the exact quantity of metal. With the flask connected to the inert gas system by a 3-way tap, reaction solvents and other reagents can be added directly to the metallic reagent.

If you need to separate the oil from a quantity of metal dispersion without placing it directly into a reaction flask, the piece of apparatus shown in Fig. 6.19 is very useful. A typical procedure is as follows:

1. After drying the filtration apparatus and cooling under a stream of inert gas, quickly weigh into it slightly more than the required quantity of the dispersion, loosely packed.

2. Add some *dry* petroleum ether to the apparatus and quickly connect a Quickfit 3-way Teflon tap to the top.

3. Open the stopcock at the bottom of the funnel and pressurize with inert gas (Fig. 6.19a). The inert gas can be introduced from a simple line which incorporates a bubbler (see Fig. 6.2a). The vent of the bubbler should be restricted to increase the pressure.

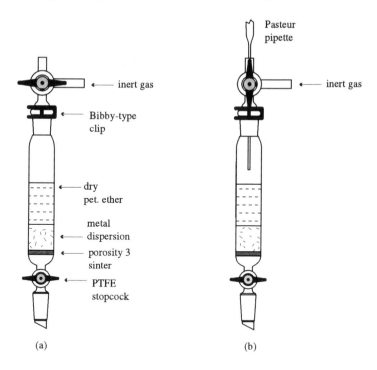

Figure 6.19

4.   Before the level of the solvent reaches the top of the dispersion, turn the 3-way tap so that argon is directed out of the vertical inlet as well as into the funnel. Then add more solvent through the vent using a Pasteur pipette (Fig. 6.19b).

5.   Repeat the washing steps until solvent flows freely, indicating that all the oil has been washed away, then close the top and bottom stopcocks.

     This procedure will leave you with finely divided metal or hydride, under argon, and this can be kept for a few hours in the sealed apparatus without deterioration. The fine powder will be very reactive and therefore great care is required in handling it. For titration of metal hydrides see Chapter 7.

Weighing reactive powdered solids

The simple procedure given below can be used for most solids, other than those which are extremely reactive with even the slightest amount of moisture or air. Examples of reagents which can be weighed by this method are: sodium hydride, potassium hydride, lithium metal, lithium aluminium hydride, finely divided metals and metal hydrides (from dispersions). A glove bag should be used for particularly reactive solids, or in cases where extreme accuracy is required.

As always, plan ahead when handling reactive materials and make sure you have a dry vessel, under inert atmosphere, into which the powder is to be weighed.

1. Remove the receiving vessel from the inert gas system and place on a balance pan under a stream of argon, provided by the simple set-up shown in Fig. 6.20. *Caution: argon is the preferred inert gas for this procedure, but because it is heavier than air, make sure the flask remains filled through-out the weighing procedure.*

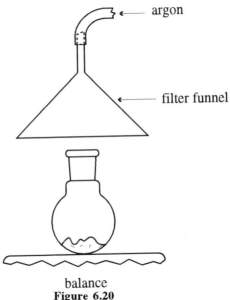

argon

filter funnel

balance
Figure 6.20

2. Keep the top of the container holding the powdered metal under the argon stream whilst removing it, and quickly weigh the required amount into the flask. **When weighing any finely powdered reactive solid avoid sifting it through the air as this can lead to a fire.**

3. Re-connect the flask to the inert gas system (preferably a double manifold), evacuate very carefully to avoid the power being disturbed, then refill with inert gas.

4. Carefully neutralize remaining traces of the metal or hydride on the apparatus using an alcohol.

The inverted filter funnel technique described above reduces contact of the reactive solid with moist air and thus lowers the chances of deterioration, and also the risk of fire. In some instances the less reactive powders, such as sodium hydride, can be safely weighed in the atmosphere if you are quick and careful. However, it is recommended that an inert atmosphere blanket is always used for weighing more reactive reagents, such as potassium hydride.

6.5 Preparation and titration of simple organometallic reagents

Grignard reagents, organolithium reagents and copper derivatives of these are commonly used as nucleophiles in organic synthesis. On the other hand, bulky lithiated amines (lithium amides) are widely used as strong bases. Some of these reagents are commercially available, but others have to be prepared. Preparation of such reagents is not difficult but a certain amount of care has to be taken. In this section we outline representative methods for preparing common organometallic reagents.

6.5.1 General considerations

The most important practical consideration when preparing organometallic reagents is that water must be rigorously excluded at all times. Most organometallic reagents can not be formed in the presence of moisture and once formed they ***normally react violently with water***. Make sure you are familiar with the principles outlined earlier in this Chapter and in Chapter 9, Sections 9.1-9.2, before attempting to prepare organometallic reagents. We recommend that you follow the following general points when preparing any organometallic reagent:

- All glassware used must be dried rigorously and assembled under an inert atmosphere.
- Coil condensers should be used if possible to avoid problems associated with water condensing on the outside of the condenser, running down to the Quickfit joint and seeping into the reaction flask.
- Quickfit joints on reaction flasks should be lined with Teflon sleeves. This is especially important when fine metal dispersions are used, which may otherwise seep through the joints and cause a potential fire hazard.

The apparatus set-ups as shown in Fig. 6.21a and Fig. 6.21b can be used for the preparation of most common organometallic reagents.

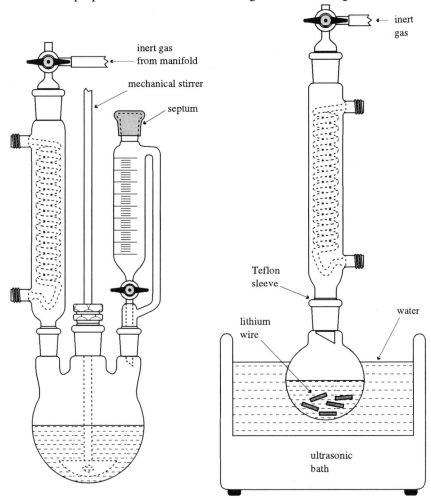

Figure 6.21

6.5.2 *Preparation of Grignard reagents (e.g. phenylmagnesium bromide)*

The apparatus shown in Fig. 6.21a, incorporating either a mechanical or magnetic stirrer, can be used for preparation of Grignard reagents: Add dry magnesium turnings (9.72g, 0.4mol) to the flask and load a solution of bromobenzene (45ml, 67.5g, 0.43mol) in dry ether (125ml) into the addition funnel. Slowly add a small amount of the bromobenzene solution (5-10ml) to the magnesium. The formation of the Grignard reagent is indicated by a cloudy appearance of the solution and the production of heat (the ether should boil gently). If the reaction does not start,[†] grind the magnesium pieces against each other with a glass rod and add a small crystal of iodine. When the initial reaction subsides add the remainder of the bromobenzene solution at such a rate that the ether boils gently. Rinse out the addition funnel with a small quantity of dry ether and add the rinsings to the flask, then stir until all the magnesium has reacted.

The solution prepared as above contains about 0.4mol of Grignard reagent. It can be reacted *in-situ* by subsequent addition of an electrophile from the addition funnel or it can be transferred to a storage container by cannulation as described in Section 6.4. Solutions of many Grignard reagents (not allyl or similar reagents) can be stored for prolonged periods under an inert atmosphere.

6.5.3 Preparation of organolithium reagents (e.g. [n]butyllithium[2])

$$\text{CH}_2=\text{CH}-\text{CH}_2-\text{Cl} \xrightarrow{\;2\,\text{Li}\;} \text{CH}_2=\text{CH}-\text{CH}_2-\text{Li} \;+\; \text{LiCl}$$

The apparatus shown in Fig. 6.21b can be used for the preparation of lithium reagents on a small scale, or an addition funnel can be incorporated

---

† Under dry conditions most organobromides react spontaneously with magnesium, but problems are occasionally encountered starting Grignard reactions and alkenyl bromides tend to be particularly problematic. Several methods have been recommended for ensuring that the reagent is prepared successfully. One method for activating the magnesium prior to use is given in Section 6.4.2 and others have been reported.[1] Addition of catalysts, such as a crystal of iodine or a few drops of 1,2-dibromoethane, is also useful. If none of these measures works try sonicating the mixture in an ultrasonic bath.

---

1. For more details about formation of Grignard reagents see: i) Y.H. Lai, *Synthesis*, 1981, 585; ii) A.G. Massey and R.E. Humphries, *Aldrichimica Acta*, 1989, **22**, 31.

2. H. Gilman, J.A. Beel, C.G. Brannen, M.W. Bullock, G.E. Dunn, and L.S. Miller, *J. Am. Chem. Soc.*, 1949, **71**, 1499.

(Fig. 6.21a) for larger scale work: Make sure the apparatus has been purged with argon (or nitrogen) then place freshly cut lithium wire[††] (5g, 0.71mol) in the 250ml reaction flask, and add dry hexanes (80ml), through the 3-way tap, *via* a syringe with a long needle. Sonicate the reaction mixture in an ultrasonic bath, then add [n]butyl chloride (37ml, 0.35mol) in dry hexanes (50ml), dropwise, *via* syringe over a period of about 10min. Reaction begins after a short induction period (*ca.* 5-10min) causing the reaction mixture to warm and producing a purple precipitate. Sonicate the mixture for a further 3h, and then leave it to settle. Once the precipitate has settled, quickly replace the condenser by a septum and carefully cannulate the organolithium solution into a clean, dry storage flask, as described in Section 6.4. If you have access to a centrifuge it is a good idea to use a centrifuge flask fitted with a septum as a storage flask. It can be centrifuged to leave any remaining precipitate as a hard layer at the bottom of the flask. Clear solutions of the reagent can then be taken from the flask by syringe, as required. As long as the solution is protected from atmospheric moisture and oxygen, it can be stored for a prolonged period prior to use. The molarity of the [n]butyllithium solution produced should be about 1.8M, and this can be checked by titration (see Section 6.5.5).

[††] The condition of the lithium used can be crucial to the success of the reaction and in many cases diethyl ether or THF are the best solvents to use. Lithium with a high sodium content should be used and it is best to clean the surface thoroughly. Lithium reagents from vinyl bromides can be particularly difficult to prepare and lithium dispersion is recommended as an alternative. Methods for separating the oil from dispersions are given in Section 6.4.5. The metal powder is very reactive and should be handled very carefully. An alternative method for preparing alkenyllithium reagents on a small scale is to react the appropriate bromide with [t]butyllithium. An equivalent of lithium halide is produced when organolithium reagents are prepared from organohalides but this can be avoided if the lithium reagent is prepared by a longer route *via* the appropriate organotin reagent.

6.5.4   Titration of lithium reagents (e.g. ^{n}butyllithium)[3]

Red-orange

1,3-Diphenyl-2-propanone *p*-toluenesulphonylhydrazone (*ca.* 200mg, measured accurately) is placed in a dry 10ml, round-bottomed flask or small Pyrex test tube fitted with a magnetic stirrer bar and septum. The system is then purged with argon (or nitrogen), and kept under a positive inert gas pressure, using a needle adapter from an inert gas line or balloon (see Fig. 9.20 or 12.3). Dry tetrahydrofuran (4ml) is then added *via* a syringe. The reaction mixture is stirred rapidly to dissolve the hydrazone, and the ^{n}butyllithium solution is added dropwise from an accurate 1ml syringe, until the colourless solution reaches an orange-red end-point. At this point the volume of ^{n}butyllithium solution added is noted, and from this the molarity of the solution is calculated using the following equation:

Molarity of ^{n}butyl lithium solution $= \dfrac{\text{amount of hydrazone (mg) x 2.64}}{\text{volume of } ^n\text{butyl lithium (}\mu\text{l)}}$

In order to obtain an accurate titre it is necessary to carry out this titration at least three times and calculate the average of the results obtained.

6.5.5   Preparation of lithium amide bases (e.g. lithium diisopropylamide)[2]

Lithium amide reagents are very sensitive to moisture and they must be prepared very carefully if they are to be used successfully. A few simple precautions should ensure that the preparation is successful:

• Titrate the butyllithium reagent immediately before using it to prepare a lithium amide.

---

3. For a discussion of the various titration methods see: J. Suffert, *J. Org. Chem.*, 1989, **54**, 509.

- Use diisopropylamine (or other amine) that has been distilled from an effective drying agent (see Section 6.3.1)
- Make sure the apparatus has been dried rigorously and purged with inert gas before you start.

Make sure you are familiar with the principles outlined in Section 9.4.1 on low temperature reactions before you start. The simple apparatus shown in Fig. 9.20 can be used for the preparation of lithium amide bases.

To prepare LDA (10mmol)[4]

Connect a dry 50ml round bottomed flask containing a dry magnetic stirrer bar to an inert gas system (e.g. as Fig. 9.20). With a steady positive flow of inert gas through the 3-way tap, add dry diisopropylamine (1.4ml, 10mmol) using a syringe, then add dry tetrahydrofuran (10ml). Set the 3-way tap so that it is open to the inert gas inlet only and cool the mixture to -78°C, in a dry ice-acetone bath. Once the flask has cooled down, add [n]butyllithium solution (5.6ml of a 1.8M solution in hexanes, 10mmol), slowly from a syringe, under a positive pressure of inert gas. Once all the [n]butyllithium has been added allow the solution to warm to 0°C for 15 min. The LDA solution can be used immediately *in-situ* or it can be allowed to warm to room temperature, sealed with a septum and stored under an inert atmosphere for future use.

For most purposes it is reasonable to assume quantitative formation of LDA using this method, consequently if the [n]butyl lithium solution has been accurately titrated, the amount of LDA formed should be known, however if problems are encountered there are procedures available to check the titre of the LDA itself.[5]

6.6 Preparation of diazomethane

Diazomethane is an extremely valuable reagent which is very easy to prepare and use.[6] However, there are certain hazards associated with the compound and its preparation that should be taken into account. The observance of a few simple safety measures allow the reagent to be prepared and used with confidence.

4. H.O. House, D.S. Crumrine, A.Y. Teranishi, and H.D. Olmstead, *J. Am. Chem. Soc.*, 1973, **95**, 3310.

5. R.E. Ireland and R.S. Meissner, *J. Org. Chem.*, 1991, **56**, 4566.

The single most valuable application of diazomethane is its reaction with carboxylic acids to provide the equivalent methyl ester, under very mild conditions. This and other reactions of the reagent have been well reviewed.[6]

6.6.1 Safety measures

There are two types of hazard associated with diazomethane: toxicity and detonation. Since the reagent is a gas, it can easily be inhaled causing serious lung disorders and a cancer risk. The precursors to diazomethane are also toxic and should also be treated with extreme care.

Diazomethane has been known to explode. Know detonation initiators to include: rough glass surfaces; alkali metals; certain drying agents such as calcium sulphate and strong light.

Fortunately, the risks highlighted can be circumvented by adhering to the following simple safety measures.

1. *Diazomethane should always be prepared and used in an efficient fume cupboard, behind a safety shield.*
2. *The detonation risk is chiefly associated with concentrated solutions or neat diazomethane, and the risk is almost completely avoided if it is always handled in dilute solution. This also reduces the health hazard.*
3. *Never use diazomethane with ground glass joints or glass apparatus with rough, broken edges and do not use anti bumping chips.*
4. *If a dry solution is required use potassium hydroxide as the drying agent.*
5. *Avoid strong light.*
6. *Do not store diazomethane, use it immediately and neutralize any excess reagent with acetic acid.*

6.6.2 Preparation of diazomethane (a dilute ethereal solution)

'Clearfit' apparatus can be used to prepare diazomethane, but a cheaper alternative is a simple one piece distillation apparatus, as shown in Fig. 6.22, which can by made by a glassblower.

6. T.H. Black, *Aldrichimica Acta*, 1983, **16**, 3.

The set-up shown in Fig. 6.21 can be used to prepare between 3 and 17mmol of diazomethane.

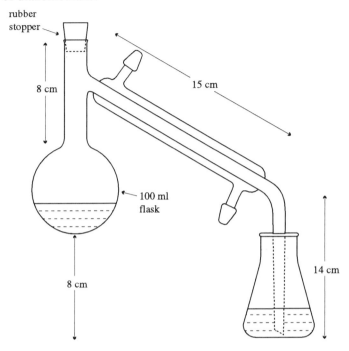

rubber
stopper

8 cm

15 cm

100 ml
flask

14 cm

8 cm

Figure 6.22

Method 1

Dissolve potassium hydroxide (1g) in water (2ml), dilute with 95% ethanol (8ml), and add to the reaction flask. Cool the flask in an ice bath, then slowly add a solution of Diazald™ (*N*-methyl-*N*-nitroso-*p*-toluenesuphonamide) in diethyl ether (15ml), using a Pasteur pipette. When all the Diazald (1g) has been added, stopper the neck of the flask, put the ice bath under the receiver and place a warm water bath (about 65°C) under the reaction flask. The yellow diazomethane-ether mixture will start to distil out of the flask. Carry-on distilling until all the yellow colour has disappeared from the reaction flask, adding a further 5ml of diethyl ether if necessary. The yellow distillate contains diazomethane (about 3.3mmol).

Method 2

To produce diazomethane (16.6mmol) use Diazald™ (5g), potassium hydroxide (5g), water (8ml), ethanol (10ml) and diethyl ether (50ml). The procedure is similar, but on this larger scale, add the Diazald solution slowly to the warm potassium hydroxide solution, while distilling off the diazomethane. A separating funnel (with a Teflon stopcock) pushed through a rubber stopper can be used for the addition.

6.6.3 General procedure for esterification of carboxylic acids

To an ice cooled solution of the carboxylic acid in ether or ethanol, add an ethereal solution of diazomethane slowly, until gas evolution stops, and a very pale yellow colour remains. Add just enough dilute acetic acid in ether to remove the yellow colour, then evaporate the mixture and purify the product as appropriate. For very acid-sensitive reactants the acetic acid work-up can be omitted, and excess diazomethane removed by bubbling nitrogen through the reaction mixture (in a good fume cupboard) until all the yellow colour has disappeared.

6.6.4 Titration of diazomethane solutions

Add an aliquot of diazomethane solution to an accurately weighed, excess quantity of benzoic acid in ether. Dilute the mixture with ether and titrate with a standard alkali solution to calculate the quantity of acid remaining.

For more information on the uses of diazomethane see reference 4.

General references for this Chapter

H.C. Brown, *Organic Synthesis via Boranes,* Wiley, New York, 1975.
D.F. Shriver and M.A. Drezdzon, *The Manipulation of Air-Sensitive Compounds,* 2nd ed., Wiley, New York, 1986.
A.L. Wayda and M.Y. Darensbourg eds., *Experimental Organometallic Chemistry, A Practicum in Synthesis and Characterization,* American Chemical Society, Washington, 1987.

CHAPTER 7

Gases

7.1 Introduction

Many organic reactions require the use of gases, either inert gases which are used to protect the reaction, or reagent gases which actually take part in the reaction. Special experimental techniques are required for handling gases and this chapter contains a summary of methods for the preparation, handling and measurement of the more commonly encountered gases.

It must be emphasized at the outset that many gases are very hazardous, either because they are toxic or because they are supplied in cylinders which contain compressed gas at very high pressures and, as a result, particular attention must be paid to safe practice in gas manipulation.

7.2 Use of gas cylinders

A large number of gases are commercially available and details about individual gases are provided in Sections 7.5 and 7.6. The purpose of this section is to describe safe methods for handling the containers in which these gases are supplied.[1,2]

Most gases are supplied in pressurized metal cylinders in sizes ranging from about 50cm in diameter and 350cm tall (lecture bottles) to the familiar size used for nitrogen (ca. 0.25m diameter by 1.5m tall). Some gases with higher boiling points are supplied at lower pressure in relatively light metal cylinders. The fittings on the cylinders vary depending on the supplier and

1. *Safe Under Pressure*, BOC Ltd, 1988.
2. *Handbook of Compressed Gases*, Compressed Gas Association, Reinhold, 1981.

on the gas hence it is very important to comply with the suppliers instructions regarding fittings and accessories.

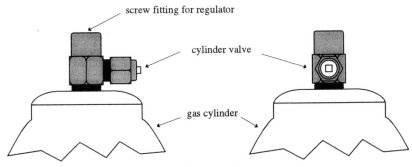

Figure 7.1

Most cylinders are fitted with a unit containing an on-off valve, and an outlet fitting (Fig. 7.1). ***This unit should never be tampered with.*** The valve fitting is a weak point in the cylinder and might be dislodged or weakened if the cylinder was dropped. Apart from releasing a potentially dangerous gas this could allow the highly pressurized gas to vent uncontrollably converting the cylinder into an extremely dangerous missile.

For this reason cylinders and lecture bottles should never be allowed to stand unsupported.

They should always be securely clamped to a bench or a wall. If they have to be moved frequently they should be supported in a sturdy metal frame, or in a trolley designed for this purpose. Cylinders should only be moved in purpose designed trolleys and should always be treated with great care.

Cylinders are generally pressurized to 175-200atm and the on/off valves provide no more control than their name suggests so cylinders must be fitted with a regulator to allow controlled delivery of the gas. The most popular regulators are the two stage types shown in Fig. 7.2.

The regulators provide a constant *outlet pressure* which can be adjusted to suit the particular application. These are often fitted with needle valves to give control over the *outlet flow rate*. A very useful accessory is a multi-way needle valve outlet which will allow more than one piece of apparatus to be connected to the cylinder at one time.

7.2.1 The procedure for fitting a regulator to a cylinder

1. Find the correct regulator for the gas in question. Different gas cylinders have different fittings (to prevent the gas from coming into contact with incompatible materials) so the appropriate regulator, fitted with the correct threaded connector, must be used. (Consult supplier's data for more detailed information.) For safety reasons it is best to choose a regulator which has a fairly low delivery pressure ($\leq$ 0-50 psi), unless a high output pressure is specifically required. (1atm = 1.01bar = 14.5psi (lbf/in^2) = 1.03 kgf/cm^2 = 760mmHg.)

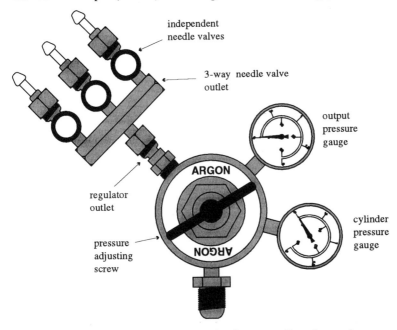

Figure 7.2 Gas cylinder regulator plus 3-way needle valve outlet

2. Remove the protective cap from the cylinder fitting and ensure that the fitting is clean and dry.
   Never use grease or Teflon tape on cylinder fittings.
3. Screw the fitting onto the cylinder and tighten firmly. A poor fit probably means that the regulator is not of the correct type. (Note that some cylinders have 'left hand' threads.)
   Never try to force the fitting.
4. Test for leaks by applying a little dilute soapy water, or a commercial leak detection solution, around the joint.

The correct procedure for controlled delivery of the gas is as follows:

1. Turn the delivery pressure adjusting screw anticlockwise until it rotates freely. The regulator is then closed.
2. Open the cylinder valve by slowly turning the spindle anticlockwise with the recommended key, until the cylinder pressure gauge shows the tank pressure. Do not open the valve any further.
3. Close the output needle valve.
4. Turn the delivery pressure adjusting screw clockwise until the required output pressure (typically 10psi) is registered on the outlet pressure gauge.
5. When the gas line is correctly attached to the apparatus (see Section 7.2) open the needle valve slowly to obtain the desired flow rate.

When the gas is no longer required the flow should be shut off and the cylinder closed by closing the regulator (rotate the pressure adjusting screw anticlockwise) and then closing the cylinder valve. The cylinder valve should always be shut when the gas is not in use.

Lecture bottles should also be fitted with a regulator and needle valve, as specified by the manufacturer. The regulators are of two types, one for corrosive and one for non-corrosive gases, and they require a Teflon or lead washer. The procedures for using lecture bottles are the same as for larger cylinders.

A number of gases are supplied in liquefied form at relatively low pressures and these generally only require the use of a flow control valve. The valves on these cylinders are generally fitted with a handwheel. Again you should consult the supplier's technical data for information on the correct fittings and handling procedures.

7.3 Handling gases

This section is concerned with general procedures for handling gases and includes a discussion of apparatus and techniques for adding them to reaction flasks. The preparation and scrubbing of some common reagent gases are described in Section 7.6. Methods for use of inert gases to protect air sensitive reactions are detailed in Chapter 9. Hydrogenation, ozonolysis, and liquid ammonia reactions are described in Chapter 14.

Two important principles must be borne in mind when carrying out a reaction involving a gas:

1. A pressure release device, such as an oil or mercury bubbler, must be attached to the apparatus in order to prevent the possibility of a dangerous pressure build-up.
2. Pressure fluctuations may result in the reaction mixture being sucked out of the reaction flask, so valves or traps must be placed in the gas line.

Reactions involving toxic gases must be carried out in an efficient fume hood.

A typical arrangement for the addition of a gas to a reaction flask is shown in Fig. 7.3. The gas is supplied from a cylinder or is prepared in another vessel and is supplied at a steady flow rate. It passes through an appropriately sized trap, such as a Dreschel bottle or an empty bubbler, before reaching the reaction flask. If a pressure reversal occurred and the contents of the flask were sucked back, the trap would prevent them from coming into contact with the cylinder or the reagents used to prepare the gas. The gas can be bubbled into the solution *via* a needle or *via* a glass tube fitted with a wide bore glass frit. Use of a frit gives better dispersion of the gas but care should be taken that the frit does not become blocked by a solid product.

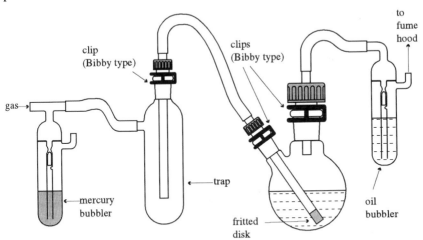

Figure 7.3

It is important to place an oil or mercury bubbler between the gas supply and the trap so that if the gas line is blocked the gas can be vented safely. Obviously the depth of oil or mercury in the bubbler must be large enough to ensure that the gas will pass through the reaction mixture rather than venting from the bubbler. With the inclusion of a bubbler as a safety device, it is advisable to put clips on all the ground glass joints, or to wrap them securely with Teflon tape, so that the gas cannot leak from the apparatus.

The type of tubing used in the line will be dictated by the nature of the gas. Rubber or PVC tubing is convenient for non-corrosive, non-toxic gases such as carbon dioxide but many reagent gases require the use of glass or resistant plastic tubing. A brief guide to the chemical resistance of some common tubing materials is provided below, and more detailed information about specific products can be obtained from manufacturers. Caution should be exercized when corrosive and toxic gases are used.

PVC (Tygon)

Flexible with low permeability but poor resistance to organic solvents and acidic gases.

Polyethylene, polypropylene

Better resistance to solvents and acids than PVC but not suitable for halogens.

Fluorocarbons (Teflon)

Poor flexibility, but excellent chemical resistance.

Natural rubber

High gas permeability and low chemical resistance.

Neoprene

Relatively good resistance to organic solvents and to acids.

Fluorocarbon rubber (Viton)

Expensive but good chemical resistance. Unsuitable for amines and ammonia.

Gas which is not absorbed can escape *via* a bubbler. Small quantities of non-toxic gases can be allowed to vent into an efficient fume-cupboard. Toxic gases should be passed through a scrubbing system and suitable procedures for various gases are described in Section 7.6. The bubbler also provides a means of monitoring the rate of uptake of the gas, and the flow rate should be adjusted so that very little is vented. For relatively insoluble gases, or for slow reactions, it may be necessary to stir or shake the reaction flask.

7.4 Measurement of gases

The addition of an accurately measured quantity of a gas is sometimes necessary, for example to obtain selective hydrogenation of a diene. Some convenient and reasonably accurate methods for measuring gases are described in this section.

7.4.1 Measuring a gas using a standardized solution

If a gas is soluble in a suitable solvent, and the concentration of the solution can be determined by a simple analytical technique, then accurately measured quantities of the gas can be dispensed by using the appropriate volume of the solution. For example solutions of the hydrogen halides in various solvents can be determined by simple acid/base titration and solutions of chlorine in carbon tetrachloride can be determined by addition of excess potassium iodide and back titration with sodium thiosulphate. Refer to textbooks of inorganic analysis for details of these methods.

7.4.2 Measurement of a gas using a gas-tight syringe

Commercially available gas-tight syringes provide the most convenient method for dispensing small volumes of gases. A syringe fitted with an on/off valve and a needle (Fig. 7.4) is repeatedly filled with the gas and then emptied again, to remove the air. It is then filled to the required volume from a gas stream, pressurized by a bubbler, and the valve is shut.

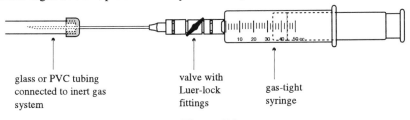

glass or PVC tubing connected to inert gas system

valve with Luer-lock fittings

gas-tight syringe

Figure 7.4

Just before use, the valve is opened to purge the needle and allow the gas to reach atmospheric pressure, and then the needle is inserted into the reaction flask. If a small volume of gas is involved (less than 100ml) the reaction vessel can be a sealed system and the gas can be added slowly from the syringe. For larger volumes the reaction flask should be fitted with a

mercury bubbler and the gas should be added at such a rate that gas does not
vent from the bubbler.

7.4.3   Measurement of a gas using a gas burette

Gas burettes are most commonly used for low pressure hydrogenations
(Chapter 14), but they can of course be used for delivering other gases thus
providing an easy method of dispensing accurately measured volumes. A
simple gas burette design is shown in Fig. 7.5. It is operated as follows:

1.   Open the double oblique tap to the vent (fume hood), and the three-way
      tap to the burette. Raise the levelling bulb to expel most of the gas
      from the burette.
2.   Turn the double oblique tap to the gas line, turn on the gas supply, and
      fill the burette slowly, lowering the levelling bulb as the burette fills.
3.   Repeat steps 1 and 2 twice more to ensure that all of the air has been
      expelled. Then fill the burette to approximately 20% more than the
      required volume.

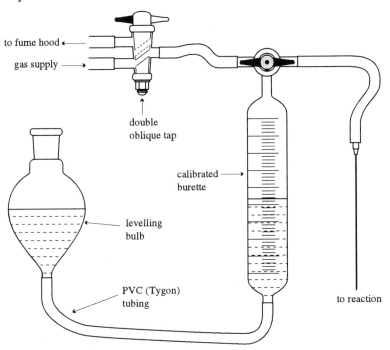

Figure 7.5

4. Open the three-way tap to the line fitted with a needle, which can be inserted into the reaction flask, and flush the tubing with gas. Turn off the gas supply.
5. Open the three-way tap to connect the burette to the line with the needle, thus allowing the gas to come to atmospheric pressure and giving the the needle a final flush. Rapidly read the volume in the burette and insert the needle into the reaction flask.
6. Raise the levelling bulb slowly to keep a slight positive pressure of the gas as the reaction proceeds. When the required volume has been added, close the three-way tap. If the gas is hazardous flush the apparatus with inert gas (steps 1-4) after use.

If gas burettes are used for a variety of different gases then some safety points should be noted. The burette should be thoroughly flushed with inert gas after each use. The tubing may need to be changed for use with different gases, and the liquid in the reservoir may also need to be changed depending on the application (aqueous copper sulphate, mercury, mineral oil, and dibutyl phthalate are often used). The gas burette method is particularly good for reactions in which the uptake of gas is slow. Gas burettes are also used for measuring the volume of gas absorbed, or evolved, in a reaction.

7.4.4   Quantitative analysis of hydride solutions using a gas burette

A useful application is the quantitative analysis of hydride solutions by adding the hydride to an excess of protic or acidic solvent, and measuring the volume of hydrogen evolved.[3]

The procedure for the analysis of $BH_3.THF$ is illustrative.

1. Place 50 ml of a 1:1 mixture of glycerol and water, and a stirring bar, in a flask and seal it with a rubber septum. Connect to the gas burette using a needle as before.
2. Open the burette to the flask and add a few millilitres of hydride solution to the flask in order to saturate the atmosphere with hydrogen.
3. Open the burette to the atmosphere and raise the levelling bulb to give a reading of approximately zero. Note this reading.
4. Turn the three-way tap so that the burette is connected only to the flask.

3. H.C. Brown, *Organic Synthesis via Boranes*, Wiley, New York, 1975.

Add an accurately measured volume of hydride solution, with rapid stirring. The volume used should be sufficient to more than half fill the gas burette you are using.

5. When hydrogen evolution has ceased, lower the levelling bulb so that the liquid levels in the burette and the reservoir are equal, and read the volume.

Calculate the molarity using the following equation.

$$\text{Molarity} = \frac{(P_a - P_s)(273)(V_h - V_a)}{((760)(T)(22.4)(V_a)}$$

P_a = Atmospheric pressure (mmHg)
P_s = Vapour pressure of the solvent (mmHg) in the resevoir at temp. T
V_h = Volume of hydrogen evolved (ml)
V_a = Volume of hydride solution added (ml)
T = Temperature (°K)

6. Repeat the measurements until reproducible results are obtained. This method should give an accuracy of better than ±5%.

7.4.5   Measurement of a gas by condensation

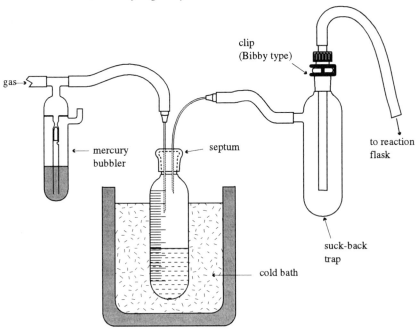

Figure 7.6

Many gases (see Appendix 2) have relatively high boiling points so they can easily be liquefied and measured in the liquid form. The gas is passed into a vessel, which is then cooled to the appropriate temperature, as shown in Fig. 7.6. The quantity of condensed gas can be measured by volume or by weight. When the required quantity has been condensed the cylinder is closed and the reaction flask is attached to the safety trap. The cooling bath is then removed and the gas allowed to distil into the reaction vessel. The mercury bubbler and safety trap serve the usual functions.

7.4.6 *Measurement of a gas using a quantitative reaction*

A specific quantity of gas can be added to a reaction if the gas is prepared using a chemical reaction of known yield. Methods and apparatus for preparing some commonly used gases are described in Section 7.6 and mention must also be made of an elegant gas generator devised by H.C. Brown, which is described in detail elsewhere.[3]

7.5 Inert gases

Nitrogen and argon are by far the most commonly used inert gases and techniques for conducting reactions under an inert atmosphere are described in Chapter 9. Both N_2 and Ar are available in various grades of purity and usually gas of 99.995% purity can be used without further purification. Attempted purification using conventional drying trains is generally counter-productive. An easy method of checking whether your inert gas line is absolutely dry is to place a flask containing some titanium tetrachloride on the manifold. Evacuate carefully for a few seconds and then open to the inert gas. If you see white fumes of titanium dioxide, you have a poor batch of gas or a leak in your gas line. Effective gas purification systems are described elsewhere.[4,5] Although it is more expensive, argon is superior to nitrogen in two important respects. It is heavier than air and therefore protects the contents of a flask more effectively, and it is completely inert whereas nitrogen does react with some materials, the most commonly

4. D.F. Shriver and M.A. Drezdzon, *The Manipulation of Air-Sensitive Compounds*, 2nd ed., Wiley, New York, 1986.

5. A.L. Wayda and M.Y. Darensbourg Eds., *Experimental Organometallic Chemistry, A Practicum in Synthesis and Characterization*, American Chemical Society, Washington, 1987.

encountered examples being lithium metal and some transition metal complexes.

7.6 Reagent gases

It may occasionally be necessary to prepare a gas, if a cylinder is not available. Methods for the preparation (and scrubbing) of some simple gases are provided in this section. Most of these preparations can be carried out in the apparatus shown in Fig. 7.7.

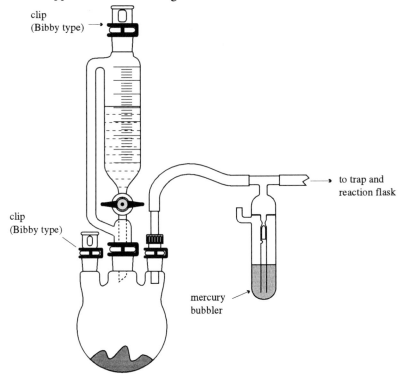

Figure 7.7

The solution in the pressure-equalizing dropping funnel is added *slowly and cautiously* to the reagent in the flask, with stirring and cooling if necessary. The gas then passes through a suck-back trap and on to the reaction flask, and any that is not absorbed passes through a scrubber. As usual, there are two bubblers, the first to prevent pressure build-up and the second to monitor the amount of gas which is not being absorbed. The rate of addition of the reagents should be adjusted so that relatively little of the gas escapes through the second bubbler. This bubbler also provides a gas-tight seal to the atmosphere which protects the reaction flask from the reagent (often water) used in the scrubber. If a toxic gas is being generated, the apparatus should be flushed with an inert gas before dismantling.

7.6.1   Gas scrubbers

Various types of scrubber may be employed but the simplest method of scrubbing relatively small quantities of water-soluble gases is to pass them through water. Two simple devices for this purpose are shown below. The first (Fig. 7.8a) is suitable only for small quantities of gas whereas the second (Fig. 7.8b) is more appropriate for larger volumes.

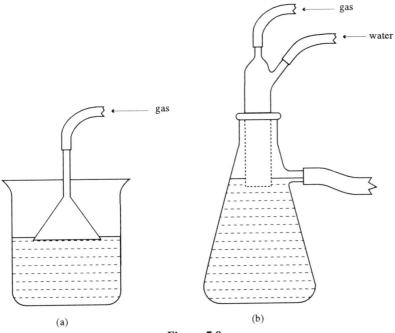

Figure 7.8

7.6.2   Methods for preparing some commonly used gasses

Methods for preparing some common gases are described below (see ref. 3 for other methods). The properties of these and other gases are listed in Appendix 2. Many of these gases are very harmful and they should only be handled in an efficient fume cupboard, taking care to prevent leaks and to scrub any excess gas.

Acetylene

Prepare by addition of water to calcium carbide and dry by passing through a column of molecular sieves. Acetylene can explode spontaneously, especially when pressurized. For this reason the gas should be generated very cautiously, the generator should be cooled, and the gas line should be fitted with a bubbler so that the pressure cannot rise above 15 psi. Dispose by venting slowly in an efficient hood.

Carbon dioxide

The simplest method is to allow solid carbon dioxide to evaporate, and the gas can be dried by passage over molecular sieves. It can also be prepared by addition of dilute hydrochloric acid to calcium carbonate. Passage over molecular sieves will remove any aqueous acid in the gas stream.

Carbon monoxide

Carbon monoxide is very toxic by inhalation: TLV 50ppm. Prepare by slow addition of anhydrous formic acid to concentrated sulphuric acid at 90-100°C (frothing tends to be a problem). One millilitre of formic acid generates 26.6 mmoles of gas. The CO is contaminated with small amounts of carbon dioxide and sulphur dioxide which may be removed by passage over potassium hydroxide or the commercial product Ascarite (sodium hydroxide on silica). Dispose of carbon monoxide by slow venting in an efficient hood.

Chlorine

Chlorine is extremely toxic and a powerful irritant: TLV 1ppm. Prepare by slow addition of concentrated hydrochloric acid to potassium permanganate (6.2ml of acid per gram of permanganate), with occasional shaking. The gas evolution slows as the reaction proceeds and warming is required to complete the reaction; 1.12g of chlorine should be formed per gram of permanganate but in our experience the yield is substantially lower than this. Chlorine may be stored in solution in carbon tetrachloride. Dispose by dissolving in water.

Diazomethane

See Chapter 6.

Hydrogen bromide

HBr is corrosive and toxic: TLV 3ppm. It can be prepared by slow addition of bromine to purified tetralin, with stirring. The small amount of bromine which passes over with the HBr can be removed by passing the gas through a trap containing pure dry tetralin. Dispose by dissolving in water.

Hydrogen chloride

Hydrogen chloride is corrosive and toxic: TLV 5ppm. Prepare by adding concentrated sulphuric acid to anhydrous ammonium chloride and dry by passage over molecular sieves. Dispose by dissolving in water.

CHAPTER 8

Vacuum Pumps

8.1 Introduction

A variety of tasks in organic chemistry require provision of a vacuum source, but different tasks require different levels of vacuum. Vacuum supplies can be loosely divided into three categories:

A *low vacuum* of between 20-50mmHg is sufficient for rotary evaporation of most solvents, filtering under vacuum, distillation of relatively volatile oils, and similar tasks.

A *medium vacuum* of about 10-12mmHg is ideal for a variety of tasks including rotary evaporation, distillations, and for serving double manifold-type inert gas lines.

A *high vacuum* of below 1mmHg is necessary for certain operations and these include removing last traces of solvent or moisture from small quantities of material and distillation of high-boiling oils. They are also ideal for serving double manifold-type inert gas lines.

8.2 House vacuum systems

Many large laboratory buildings have a large central vacuum pump which serves a house vacuum system. There are great variations in the effectiveness of such systems, depending on the type of pump employed, the number of users and various other factors. A good house vacuum system will pull about 50mmHg and can be very useful for driving rotary evaporators, filtrations and simple vacuum distillations. The advantage of the system is ease of use, but the vacuum can fluctuate quite considerably depending on how many people are using it.

8.3 Medium vacuum pumps

8.3.1 Water aspirators

Water aspirators are widely used to provide a modest vacuum of about 10-12mmHg using a good mains cold water supply. The advantages of a water aspirator are that it is cheap and simple to operate. Its main disadvantage is that it is prone to 'suck-back', which can leave your apparatus full of water! This occurs if the water pressure drops after a good vacuum has built up in the apparatus. The same thing will happen if the tap is turned off while the system is still under vacuum. For this reason the tap should always be kept full on, and an air inlet should be provided by means of a 3-way tap. A simple trap should also be incorporated, which will give you some time to disconnect the apparatus if suck-back occurs (Fig. 8.1). A water trap will not prevent suck-back altogether and hence it is not advisable to leave apparatus unattended while connected to an aspirator. This type of pump is not suitable for connecting to a manifold-type inert gas line. Other disadvantages of the water aspirator are that it is quite noisy and uses a considerable volume of water if it is left on for prolonged periods.

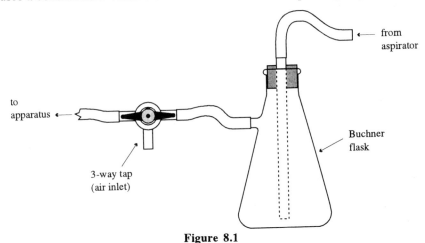

Figure 8.1

8.3.2 Electric diaphragm pumps

Various oil-free electric diaphragm pumps can be used to provide vacuum levels down to about 10mmHg. Some models are Teflon lined (e.g. KNF, Vacuubrand, Divac) and designed to withstand solvent vapours, and these are especially useful. They can be used instead of water aspirators for medium pressure distillations, evacuating rotary evaporators and a range of other tasks. The problem of 'suck-back' is circumvented using a diaphragm pump and they do not use any water. For many tasks they can also be used instead of high vacuum oil pumps, with the advantage that liquid nitrogen trapping systems are not required. For most purposes they can be used as the vacuum source for double manifold type inert gas lines (described in Chapter 4, Section 4.4.3, Chapter 6 and Chapter 9).

For some tasks very cheap and simple fish tank-type diaphragm pumps are useful. These will produce a vacuum of about 300mmHg and can also be used to provide compressed air for pressurizing chromatography columns (see Chapter 11).

8.4 High vacuum pumps

8.4.1 Rotary oil pumps

Rotary oil pumps will provide a reliable vacuum down to about 0.01mmHg (at the apparatus) and are thus one of the most valuable pieces of equipment in the lab. Ideally every research worker should have his/her own high vacuum pump, but in many cases cost prohibits this and pumps are shared. If a pump is shared, it is common to have it mounted on a trolley which has all the ancillary devices (traps, gauge etc.) mounted on it. Alternatively, a shared pump may be fixed in one place, but attached to a communal manifold, distillation set-up or other piece of apparatus. A good two-stage pump is suitable for most high vacuum requirements in an organic chemistry lab.

Solvent traps

A high vacuum pump must be fitted with efficient solvent traps to prevent the pump oil from becoming contaminated. The traps are simply cold finger condensers incorporated into the vacuum line before it enters the pump. It is preferable to use a double condenser system, as shown in Fig. 8.2, with the traps being cooled by either liquid nitrogen, or solid CO_2-acetone. If a single condenser system is used liquid nitrogen must be used as the coolant because CO_2-acetone will be ineffective.

If the trapping system incorporates a cone joint, it can be connected directly to a manifold such as that shown in Fig. 4.1 (Chapter 4).

Alternatively, a piece of high vacuum tubing can be used to connect the traps to a distillation apparatus, a double manifold (Fig. 4.4, Chapter 4), or any other piece of equipment.

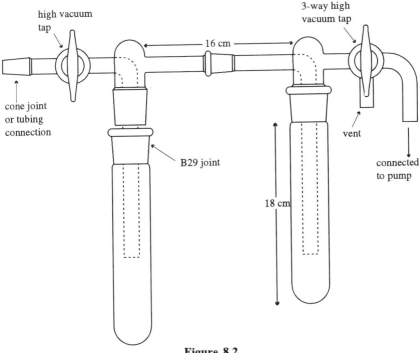

Figure 8.2

Operating a high vacuum pump

Do not immerse the solvent traps in liquid nitrogen unless they are evacuated, as this will cause liquid oxygen to condense. The result of this can be a violent explosion, caused by re-vaporization or by oxidation of organic solvents etc.

Before switching the pump on check that the traps are empty. Then with the 2-way tap closed and the 3-way tap open between the pump and the traps, but closed to the vent, turn the pump on. Next immerse the traps in Dewars filled with liquid nitrogen and check the vacuum by connecting a vacuum gauge to the 3-way tap. If the vacuum is satisfactory (<1mm), the 2-way tap can be opened, to evacuate the apparatus which is connected to it. The vacuum should then be checked again to make sure there are no leaks in the system.

When you want to switch the pump off, close the 2-way tap, turn the 3-way tap to vent through the pump (*not into the traps*), then turn the pump off. Remove the liquid nitrogen Dewars from the traps immediately after switching the pump off and then vent air into the traps using the 3-way tap. It is very important to vent the pump before turning it off otherwise pump oil may suck back into the traps.

Care of high vacuum pumps

High vacuum pumps are very expensive items, but they will give reliable service for many years if they are treated properly. For an organic chemist to work efficiently he/she must be able to rely on pumps working properly and this will only be the case if they are treated with care. The main reason for deterioration of vacuum pumps is contamination of the oil with volatile organic compounds and acidic vapours such as HCl or HBr. To avoid this problem, you should be very conscientious about keeping the coolant well topped-up around the traps, and emptying them regularly. No matter how careful you are, some solvent vapour will always find its way into the pump, and for this reason the oil should be changed regularly, at least once a month for a pump which is used every day.

8.4.2 Vapour diffusion pumps

Occasionally a distillation will require a higher vacuum than that produced by a rotary pump. Higher vacuums are normally obtained by employing a mercury or oil vapour diffusion pump, in conjunction with a rotary pump. Oil vapour pumps are most commonly used today and various models are now commercially available which will produce a vacuum of about 10^{-5}mmHg. These pumps are only used occasionally but it is very useful to have one shared between a large group or section.

8.5 Pressure measurement and regulation

It is very important to be able to measure the pressure in a vacuum system, particularly when carrying out a distillation. For low vacuum measurement a simple manometer, such as that shown in Fig. 8.3a, is commonly used and the pressure is taken by subtracting the heights of the mercury levels. Dial gauges are also useful for in-line measurements and they are particularly valuable when used with rotary evaporators. For high vacuum measurement a McLeod gauge, with a range of 1 - 0.001mmHg, is most commonly used (Fig. 8.3b). The gauge is rotated into the vertical position to read the vacuum, but must be turned back to the horizontal position and back again to vertical, in order to read a change in pressure. Electronic Pirani gauges are also available for measuring high vacuum.

Quite a common requirement is for a distillation pressure somewhat higher than that delivered by the high vacuum pump. Achieving this is no simple matter because a very accurate leak must be provided in order to maintain a constant vacuum. There are various devices available, one of the simplest being an accurate needle valve (such as an Edwards LV5) which will be suitable for most distillation purposes.

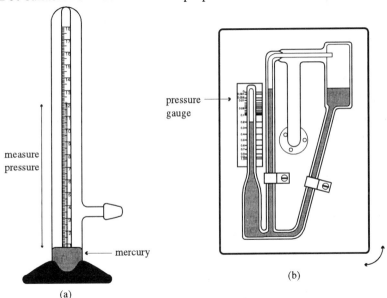

Figure 8.3

8.5.1 Units of pressure (vacuum) measurement

There is often confusion about different units of pressure measurement, but the way they relate to one another is quite simple. The most common units are mm Hg, and 1mm Hg is the same as 1Torr. The other common scale is mbar which relates to atmospheric pressure, 1 bar = 1 atmosphere = 760mm. Thus:

$$1mbar = 0.76mmHg = 0.76Torr$$

CHAPTER 9

Carrying Out The Reaction

9.1 Introduction

This is perhaps the most important single chapter in this book. In order to be certain of the results of a particular experiment, and to ensure reproducibility, it is important that the appropriate precautions are taken, and that proper preparations are made. Many common organic reactions involve the use of air sensitive reagents, which requires the reaction to be carried out under inert, anhydrous conditions. These conditions are also used for reactions which are sensitive to the presence of water. Accordingly most of this chapter is concerned with the use of air sensitive reagents, since the same basic techniques can be used for almost all types of reactions likely to be encountered.

First of all, before attempting the reaction it is important to ensure that you have the necessary glassware, apparatus, reagents, and experimental procedure (including work up and a tlc system which will allow you to follow the reaction). Do not forget to check on the safety aspects of the chemicals (including solvents), of the likely reaction products, and of the procedures which you will be using. As alluded to above, most reactions will require monitoring of some kind (tlc, gc, hplc, etc.) and it is important to ensure that your chosen system is suitable for the starting material, and give some consideration as to where the expected product is likely to be found (will the polarity increase or decrease on converting starting material to product?). Make certain that you allow yourself plenty of time to complete the reaction, or at least to monitor it in the early stages if it is likely to be a lengthy experiment. Finally it is most important that the reaction is written up carefully as you go along, rather than writing it up "properly" a few days later for the reasons given in the section on keeping your laboratory notebook. Remember, it is important to write up experiments which have "failed" as carefully as those which are successful. If this is done the reason for "failure" might be apparent to you at a later stage, and in any case this detailed information is likely to be of real value to someone following up your work.

9.2 Reactions with air sensitive reagents

9.2.1 Introduction

Many commonly used organic reagents are extremely reactive towards water and/or oxygen. Some examples are: alkyllithium reagents, Grignard reagents, organoboranes, metal hydrides, organoaluminium compounds, cuprates, titanium tetrachloride, dryed solvents, etc. Some of these reagents are available commercially and others are prepared *in situ* in either case they are most conveniently handled in solution. Although extreme care must be taken to exclude air and moisture when using these reagents, you should find them easy to handle once you become familiar with some simple techniques. General procedures for handling air sensitive reagents are described in Chapter 6 and the appropriate parts of that chapter should be consulted in conjunction with this section.

9.2.2 Preparing to carry out a reaction under inert conditions

If you already have permanent inert gas lines in the lab as suggested in Chapter 4, then carrying out a reaction under inert conditions should be routine. However, it is very important that you do not expose the reaction to the atmosphere at any stage. The techniques for ensuring this become second nature once you gain experience; but forward planning is always essential so that everything is in the right place at the right time and under an inert atmosphere. Therefore, before you start, think very carefully about how you intend to carry out each operation of your reaction sequence, and list the equipment you will need. All equipment needs to be dried and cooled in advance and it is particularly important to have enough syringes available. Indeed, you should always have some spare dry syringes in case one gets jammed, contaminated or broken.

9.2.3 Drying and assembling glassware

It is assumed that the glassware has been cleaned before use and it is also important to check glassware, particularly flasks, for cracks and scratches especially if the glassware is going to be evacuated. If you are in doubt as to the safety of a particular item of glassware do not use it, and do not forget to get it repaired by the glassblower or to dispose of it. Do not leave it around for someone else to discover its weakness. Under normal conditions glassware has a thin film of water adsorbed to its surfaces which must be removed before moisture-sensitive reagents, dried solvents, or starting materials are allowed to come into contact with it. There are two basic methods for doing this. The first method is to heat the glassware in an oven at a temperature above 125°C for at least 6h, then quickly assemble it whilst hot and cool under a stream of dry inert gas. The second method is to assemble the glassware, evacuate it by connecting it to a double manifold/high vacuum pump, then heat the whole set-up with either a Bunsen or heat-gun. The two-way tap on the manifold is then switched so that the apparatus filled with argon (or nitrogen) while it cools, leaving the apparatus set up and ready for use. This procedure can save a good deal of time but, *because of the dangers involved in heating glassware under vacuum, this procedure is best reserved for small-scale reactions.*

Glassware and normal syringes can be dried in an oven then cooled in a desiccator over P_2O_5 or self-indicating silica gel, but it is not advisable to heat microlitre syringes and these should be dried under vacuum in a desiccator.

If you are going to use magnetic stirring, do not forget to include a stirrer bar before the apparatus is assembled and before the drying procedure is carried out, since the stirrer bar will also be covered in a film of moisture and adding it later is likely to introduce air and moisture into the system.

It is not advisable to use grease for glassware assembly, since this will eventually find its way into your reaction product. If you have to use it, always use the minimum quantity possible and remove it carefully from the joint immediately after the reaction has been quenched. This can be done using a tissue dampened with chloroform or other halogenated solvent. It is important to do this before addition of an extraction solvent, or before pouring the reaction mixture through the joint. Rather than use grease, it is much better to use Teflon tape or sleeves, since these are at least as efficient, do not contaminate your product, and in the case of sleeves they are reusable. It is also advisable to seal the outside of the joint by wrapping Teflon tape around to cover the 'join'. This will help to prevent any condensation on the outside of the glassware from finding its way into the reaction. Joint clips (e.g. Bibby clips) should be used on all accessible joints.

9.2.4 Typical reaction set-ups using a double manifold

A double manifold (Fig. 9.1) is an extremely useful piece of apparatus, especially for carrying out reactions under inert atmosphere and its installation is described in more detail in Chapter 4. One barrel of the manifold is connected to a high vacuum pump (or house vacuum system) and another to a cylinder of dry inert gas. Its main advantage is that, simply by turning the two-way tap on one of the outlets, any vessel connected to that outlet can be switched instantaneously from vacuum to inert gas atmosphere or *vice versa*. Several reactions can also be set up at once, the only limit being the number of outlets. At the end of the argon (or nitrogen) line there is a bubbler which prevents air from being sucked back into the system, and provides a convenient means of monitoring the gas flow rate.

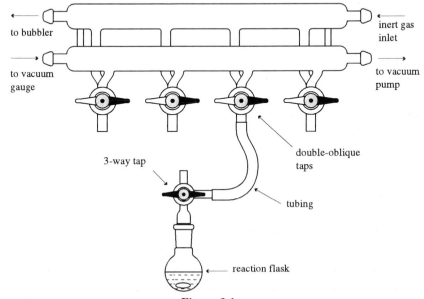

Figure 9.1

A fast flow is used when filling a system, but once the reaction is set up, one bubble every 5-10s is adequate. A piece of apparatus which is very useful when used in conjunction with a double manifold is a Quickfit 3-way tap, which is also described in more detail in Chapter 4.

9.2.5 Basic procedure for inert atmosphere reactions

A typical simple reaction set-up using a manifold is shown in Fig. 9.1 and the following general procedure is applicable to a wide range of reactions. However, this is only intended to be a general guide and can easily be modified to suit various requirements. Some of these modifications are discussed later in this chapter.

Before starting, decide on the order for addition of reagents and if possible arrange this so that any solid reagent is added first. If this is not possible try to add the solid as a solution, using the technique described later. If a reagent really does have to be added as a solid, this can be done successfully and again a method will be described to do this.

Make sure before you start that you have all the *dry* syringes, needles, and any other ancillary items, and that you are familiar with the operation of the manifold and the 3-way Quickfit tap (Fig. 9.2).

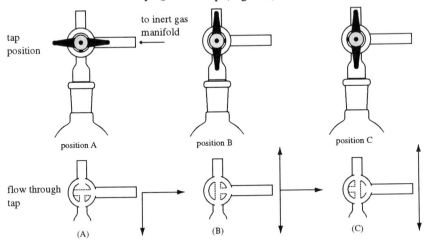

Figure 9.2 Flow through a 3-way tap relative to tap position.

1. Set up and dry the system

Connect the reaction flask *including stirrer bar* to the manifold (Fig. 9.1), set the 2-way tap on the manifold and the 3-way tap on the reaction flask (position A, Fig. 9.2a) so that the system is evacuated, and heat the glassware with a heat gun (or Bunsen) for a few minutes. Then, with inert gas flowing through the bubbler at a rapid rate, turn the manifold tap *slowly* to introduce the inert gas. Keep an eye on the bubbler while you are switching to the inert gas supply and operate the tap carefully to avoid 'suck-back'. Finally leave the apparatus to cool.

2. *Add initial reactants*

    Solid reactants (or heavy oils) can be weighed into the reaction flask at this stage. With the inert gas flow at a reasonably high level, remove the flask from the system and quickly weigh the reagent into it (for reactive solids see Chapter 6). Then re-attach the flask to the 3-way tap and switch the manifold tap to evacuate. If there are traces of solvent on the reactant these can be removed by leaving the flask under vacuum for an extended period, otherwise the 2-way tap can be switched back to argon straight away leaving the reactant in the flask under inert atmosphere.

3. *Add solvent*

    The best way to add solvent to the system, and avoid breaking the inert atmosphere, is to use a syringe. It is best to syringe the solvent straight from the collecting head of a still, which is itself under inert atmosphere (see Chapter 5), but the solvent might also be taken from a bottle or flask under inert atmosphere (see Chapter 6). In either case, once you have a syringe loaded with solvent, open the 3-way tap to position B (Fig. 9.2b), keeping an inert gas flow *which is rapid enough to maintain bubbling*, and add the solvent. Then turn 3-way tap back to position A (Fig. 9.2a) and reduce gas pressure to maintain steady bubbling. For large scale reactions solvents can be added by disconnecting the flask, adding the solvent quickly and re-assembling, but this inevitably leads to air entering the system. With higher boiling solvents the air can be removed by sequentially partially evacuating and refilling with inert gas several times, using the 2-way manifold tap.

4. *Cool the flask*

    At this stage the reaction flask can be immersed in a cooling bath and stirred with a magnetic stirrer. *Never cool a flask before it is under the inert atmosphere as this causes moisture to condense.*

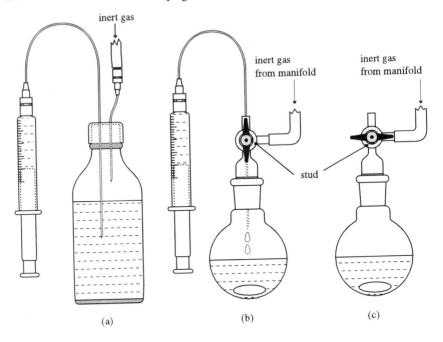

Figure 9.3

5. *Add reagents*

    It is most commonly the reagents used in the reaction which are air
    and/or moisture sensitive and they should be added with great care (see
    Chapter 6 for more detail). After pressurizing the reagent container
    with inert gas, syringe out the required quantity (Fig. 9.3a). Increase
    the inert gas flow through the manifold so that rapid bubbling is
    maintained as the 3-way tap is opened to position B, then insert the
    syringe through the tap to deliver the reagent (Fig. 9.3b). When all the
    reagent has been added turn the 3-way tap back to position A and
    reduce the gas flow so that there is a bubble every few seconds (Fig.
    9.3c). Any further reagents are added in the same way and the reaction
    is left to proceed. To add a cooled reagent it is advisable to use a
    cannula made out of Teflon tubing (see Chapter 6, Section 6.4.3 for
    more detail).

6. *Reaction monitoring*

    The reaction can easily be monitored by tlc, gc, etc. Again increase the
    inert gas flow through the manifold so that rapid bubbling is
    maintained as the 3-way tap is opened to position B, then insert a

drawn piece of glass tubing or Pasteur pipette through the tap to remove a drop of the mixture.

7. *Work-up*

    Once the reaction is complete it can be worked up as normal (see Chapter 10). However, when there is the likelihood of residual reactive reagents such as organometallics, be very careful when adding aqueous solutions. It is normally safer to keep the mixture under an inert atmosphere whilst the solution is added.

9.2.6 *Modifications to basic procedure*

1. Reactions at elevated temperatures

When a reaction needs to be heated or when there is any possibility that it

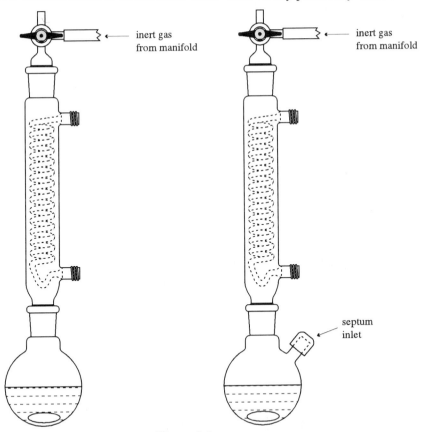

Figure 9.4

might be exothermic, a condenser should be incorporated into the reaction
set-up. Typical set-ups for reactions which are to be heated are shown in
Fig. 9.4. Normal Liebig condensers have the water jacket next to the outer
surface and there is always the possibility of atmospheric moisture
condensing on the outside, running down to the ground glass joint, and
seeping into the reaction flask. Teflon sleeved joints should prevent water
getting into the reaction vessel, but a better method is to use coil type-
condensers for reactions under inert conditions. Since the cooling surfaces
are inside the apparatus, the problem of outside condensation is alleviated.
The procedure for carrying out the reaction is exactly as described in the
previous section, except that the reaction is heated instead of being cooled.

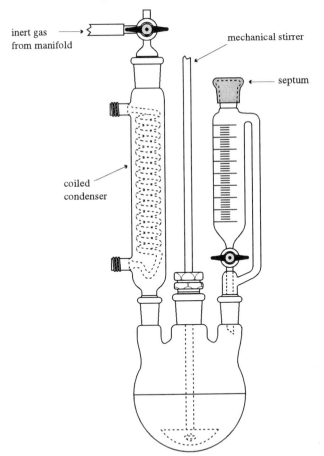

Figure 9.5

Reagents can be added through the 3-way tap, provided a syringe with a long needle is used, or as an alternative a flask with a side arm fitted with a septum can be used. *Whilst the temperature of a reaction is changing the bubbler of the system should be checked to make sure there is a constant inert gas pressure.*

2. Slow addition of reagents and large scale work

When slow addition of an air-sensitive reagent is required it is best to incorporate a pressure equalizing addition funnel into the apparatus, and for

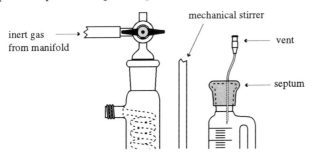

(a) Cooling apparatus under inert atmosphere

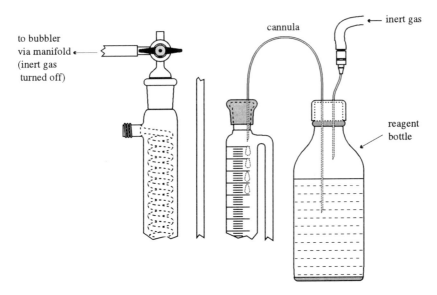

(b) Filling dropping funnel using cannula

Figure 9.6

large scale work this is always best. The apparatus can be set up with or
without a condenser, can be stirred either magnetically or mechanically. A
typical arrangement is shown in Fig. 9.5. Where very careful slow addition
is required, particular on a small scale, the use of a syringe pump is strongly
advised. These are mechanical devices which drive the plunger of a syringe
very slowly into the barrel. To dry this larger, more elaborate set-up, it is
best to oven-dry the equipment first, assemble it hot and leave to cool under
a stream of inert gas. This can be done by connecting to inert gas from the
manifold, with a short needle through the septum acting as a vent (Fig.
9.6a). Once the apparatus is cool, reactants can be added as described
above. A syringe can be used to add small quantities of reagent to the
addition funnel. For larger quantities (>50ml) the cannulation technique
described in Chapter 6 for bulk transfer of liquids is best, and if the funnel is
graduated the reagent can be added straight from the reagent bottle (Fig.
9.6b). If a mechanical stirrer is used a Teflon sealed guide should be used
for the stirrer rod. The reaction is carried out exactly as before.

3. Addition of a reagent prepared separately, or of a solid

When a solid reagent is to be added to a reaction set-up which is already
under inert conditions, it is best to add it in solution. To do this, weigh the
solid into a dry flask, fit the flask with a 3-way tap, connect to a separate
outlet on the manifold, and evacuate. Then switch the manifold tap to

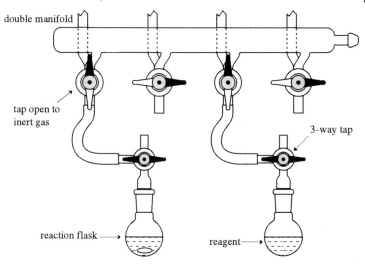

Figure 9.7

introduce inert gas, and with the gas flow high enough to maintain rapid bubbling at the bubbler, open the 3-way tap to position B and syringe in the required dry solvent. With the 3-way tap turned back to position A, the solution is under inert atmosphere and can be kept indefinitely (Fig. 9.7). To transfer the reagent, make sure that the inert gas flow is high enough to maintain rapid bubbling whilst the 3-way tap on the reagent vessel is opened to position B. Then syringe out the solution, turn the tap back to position A, open the tap on the reaction flask to position B and add the reagent. Alternatively, a cannula can be used to transfer the reagent (see below).

A similar procedure can be used when a reagent has to be prepared under inert conditions for use in the reaction. This is quite a common occurrence, for example when a Grignard or alkyllithium reagent has to be prepared, then added to a reaction mixture. In this case the reaction flask assembly is set up attached to one outlet of the manifold, while the reagent is prepared in an appropriate set-up at another outlet. The reagent can easily be transferred to the reaction flask using a syringe.

For large scale transfers, or for the transfer of a pre-cooled solution, a cannula can be used. To use a cannula, first fit the 3-way tap of the reagent flask with a septum, turn the tap to the closed position and disconnect from the manifold. Then pressurize the flask with a separate inert gas supply,

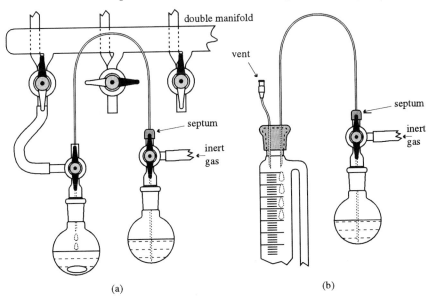

(a)   (b)

Figure 9.8

turn the 3-way tap to position B, insert a cannula, and transfer the solution either to the reaction flask directly (Fig. 9.8a), or to an addition funnel (Fig. 9.8b). For more information about cannulation and adjustment of inert gas pressures see Chapter 6, Section 6.4.2.

4. Direct addition of a solid to a reaction under inert atmosphere

Although it is much easier to add solids in solution, on some occasions a solid will have to be added directly. With the apparatus connected to a manifold and with argon as the inert gas, the procedure is often quite simple. A reaction flask with a stoppered side arm should be used and, after first increasing the argon flow to give very rapid bubbling, the stopper can be removed to add the solid reactant. If this procedure is used, great care should be taken. The stopper should be removed for the minimum time required to add the reagent, and you should make sure that it is not added at a rate which might lead to the reaction going out of control. Alternatively, an inverted funnel (see Chapter 6) can be used to blanket the apparatus with an inert atmosphere to protect the reaction mixture.

There are alternative arrangements for adding solids, but none are universal. One method is to attach a bent tube containing the solid to the side arm of the flask, and twist the tube to allow the compound to pour into the reaction flask (Fig. 9.9).

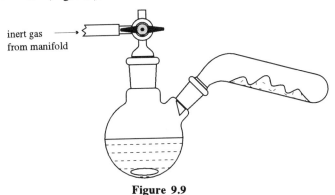

inert gas
from manifold

Figure 9.9

9.2.7 Use of balloons for holding an inert atmosphere

Although we recommend using the double manifold techniques described above, reactions can also be kept under inert atmosphere by using a balloon filled with inert gas (Fig. 9.10). This can prove particularly useful if a double manifold system is not available. To do this you first fill the balloon with the inert gas from a cylinder, then attach it to the reaction system using either a needle/septum or 3-way tap, taking care to ensure that there are no leaks in the system.

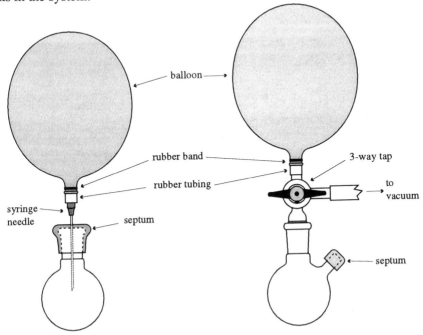

Figure 9.10

In the case of the needle/septum attachment, any air contained in the reaction flask is then flushed out by using another needle to vent the system (Fig. 9.11). After the reaction flask has been thoroughly flushed with the inert gas, the extra needle is removed and the reaction flask is ready for use. The balloon keeps the whole system under a positive pressure of the inert gas, and also allows liquid materials to be added to, or removed from, the flask *via* syringe insertion through the septum.

Generally argon is preferred when using this technique because it is more dense than air, and will fill the reaction flask, pushing out any air more effectively than would nitrogen. This technique can also be employed when syringing air-sensitive materials from bottles or containers.

With the three-way tap attachment, a vacuum line can be connected (Fig. 9.10). The system can then be purged by sequentially evacuating, then filling with the inert gas from the balloon. This is a more effective procedure than simply flushing out the flask and so both nitrogen and argon can be used with this set-up. Liquid materials can then be added to the flask by syringe *via* the septum inlet as before.

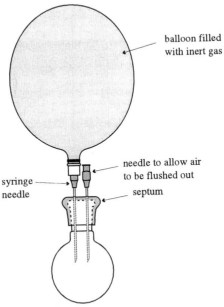

Figure 9.11

Since balloons are perishable it is often advisable to use a 'double balloon system'. This is simply two balloons, one inside the other, allowing the inert atmosphere to be maintained even if one bursts.

How to make a balloon attachment

Balloon attachments are easily constructed as outlined in Fig. 9.12. First the balloon (or double balloon) is pushed over a piece of rubber tubing which is about the same diameter as the balloon neck. This is secured in place using a piece of wire, elastic band, or Parafilm. The open end of the tubing can then be attached to a three-way tap, or to a needle-tubing adapter and needle as shown. It is advisable to seal all joints with Parafilm in order to ensure that there are no leaks.

Setting up a reaction

Once the balloon attachment has been constructed, the reaction system is simply set up as outlined in Fig. 9.10.

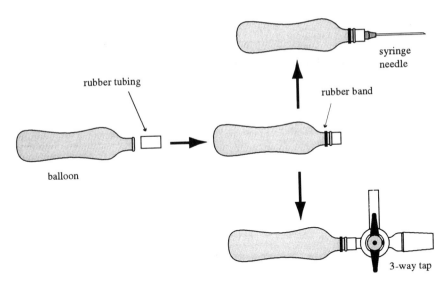

Figure 9.12

If the reaction is to be heated, a condenser will need to be fitted, in this case it may be necessary to use a two necked flask so that materials can be added to the flask. In such cases the balloon should be placed at the top of the condenser in order to prevent volatile liquids reaching it.

All joints should be sealed with Parafilm or Teflon tape, the latter being especially useful if the system is to be heated (Parafilm melts!).

9.2.8 The use of a 'spaghetti' tubing manifold

The main drawback with balloons is that they have a tendency to burst and this can have grave consequences, especially if you are working on a small scale. A spaghetti tubing manifold (Fig. 9.13) will provide a similar low pressure inert gas source, but is much more reliable. It is particularly useful if you need to set up several small scale reactions running in parallel to one another. This is often the case if you need to optimize the conditions for a reaction.

The reaction is set up in exactly the same way as described in the previous section for use of balloons. With either type of set-up a tlc sample can easily be removed by pushing a syringe needle through the septum then inserting the tlc capillary through it. When using the 'spaghetti' manifold it is often convenient to add reagents, or transfer them, by cannulation using a Teflon cannula (Fig. 9.14, see also Chapter 6, Section 6.4.2).

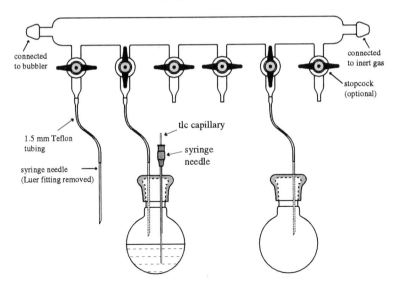

connected
to bubbler

connected
to inert gas

stopcock
(optional)

tlc capillary

1.5 mm Teflon
tubing

syringe
needle

syringe needle
(Luer fitting removed)

Figure 9.13

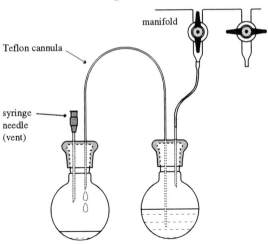

manifold

Teflon cannula

syringe
needle
(vent)

Figure 9.14

9.3 Reaction monitoring

Most people who are new recruits to the research labs learned their basic skills for carrying out reactions in an undergraduate laboratory. Inevitably, most of the organic chemistry undertaken in these lab classes involves following 'recipes', which have been well tried and tested. Therefore the conditions and the time taken for the reactions to reach completion are well established, and work-up can be carried out after a pre-set time. Unfortunately, the idea that you can guess the time it takes for a reaction to reach completion is a very bad habit to carry over into a research environment.

Every reaction you carry out should be monitored, and one of the first things you should do before starting any reaction is to decide on a suitable method for monitoring its progress. Even if you are following a literature procedure, reaction monitoring is still essential and it will usually save you time as well as giving you confidence about what is happening. Carrying out a reaction without monitoring its progress is like trying to thread a needle with your eyes closed!

The simplest and most universal method of reaction monitoring is thin layer chromatography (tlc) and this will be discussed first of all, but it is not *always* the best or only method, and sometimes you may have to use a little ingenuity to find an appropriate reaction monitoring technique.

9.3.1 Thin layer chromatography (tlc)

Tlc is a simple, but extremely powerful analytical tool. However it may take a little time before your expertise reaches a consistently high level, since a certain amount of intuition is always involved in choosing the appropriate solvent system, spotting the correct amount of sample, etc. Once you have gained experience and confidence in the use of tlc, you will find it extremely useful for a variety of purposes.

The main uses of tlc

1. Tlc is normally the simplest and quickest way to monitor a reaction and the reaction mixture should be chromatographed against starting materials (and a co-spot). This allows you to follow how the reaction is progressing, and to assess when is the best time to work it up. In all cases a record of the tlc should be made in your lab book (see Chapter 2).
2. Tlc can be used to indicate the identity of a compound, by comparing the unknown sample with a known material. In general each substance is spotted separately and also together (co-spot). Caution should be applied as co-running on tlc is *not* definitive proof of identity. Of course, substances that do not co-run are definitely not the same.

3. Tlc usually gives a good indication of the purity of a substance. Diastereoisomers can usually (but not always), be separated.
4. For flash chromatography (see Chapter 11), tlc is first used to determine the solvent system and quantity of silica required, and secondly to monitor the column fractions.

Tlc plates

There are two main types of coating for tlc plates: silica and alumina. Silica plates are most commonly used, they are slightly acidic and are suitable for running a broad range of compounds. Most of the information in this section refers directly to silica plates, but the same principles apply when using alumina plates. Alumina plates are slightly basic and are commonly used when a basic compound will not run very well on silica.

The most common tlc plates have either a glass or plastic backing coated with a thin layer of silica, which contains a binding agent to keep it bonded to the backing. Although tlc plates can be home-made, most people prefer to use commercial ones as they give consistent results. For analytical purposes plates with a 0.25mm layer of silica are normally used. They are available in a variety of sizes, although 5 x 20cm is probably the most convenient size. The chromatographs are run along the 5cm length, cut to an appropriate width. This normally gives adequate resolution, with a very short running time. Plastic backed plates are sometimes cheaper and can easily be cut into strips with scissors, but glass plates seem to give better resolution and can be heated more vigorously for visualization purposes.

To cut a glass tlc plate, place it face down on a clean piece of paper, hold a ruler firmly along the proposed cut and draw a *sharp* diamond glass cutter along the line *once only*. Then holding the plate with the forefinger and thumb of each hand, on either side of the score line snap the plate along the line. With practice, and a good glass cutter, you should be able to cut plates down to about 1.5cm without ever spoiling them.

The procedure for running a tlc

1. Cut a tlc plate which is 5cm long, and wide enough so that about 0.5cm can be left between each spot (obviously the width of the plate is dependent upon the number of spots to be run on it).

2. Make a tlc spotter by drawing out a Pasteur pipette or melting point tube to about 0.5mm using a micro burner. The spotter can be used many times provided you wash it with clean solvent inbetween runs.

3. If the substance to be analyzed is not already in solution, take it up in a volatile solvent as a ca. 1-2% solution. A non-polar solvent is preferable but dichloromethane is often used as a universal solvent for tlc samples. Reaction mixture solutions can be diluted down if necessary, and with experience making samples to a reasonable strength becomes intuitive.

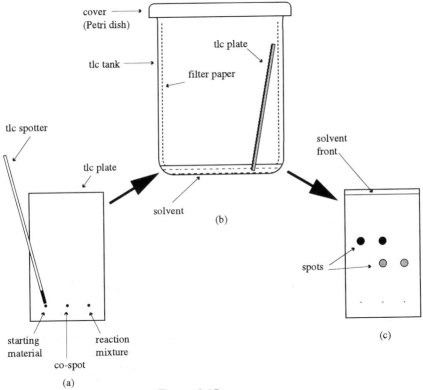

Figure 9.15

4. Using the spotter, spot a small amount of each solution about 0.5cm from the bottom of the plate, leaving a similar distance between each spot. The spots should be kept as small as possible, and you should take care to make sure that each of the spots is the same distance from the bottom of the plate (Fig. 9.15a).

The absolute distance which a compound runs up a tlc plate is extremely variable, depending on the exact conditions under which the plate was run. It is therefore much more informative to run comparative tlcs. When analyzing a reaction mixture, it should always be run in comparison with the starting material and a mixed spot should also be run. This is very important because it often enables you to distinguish between compounds which run in almost identical positions.

5.   Place the plate upright in a tank lined with a filter paper and containing the chosen solvent system (to a depth of ca. 0.3cm as shown in Fig. 9.15b). Allow the solvent to creep up the tlc plate until it is ca. 0.3cm from the top, then remove it, and mark the level of the solvent front (Fig. 9.15c). A 150ml lipless beaker makes a convenient tlc tank and a petri dish can be used as a cover.

Detecting the spots

The three general ways to visualize spots on tlc plates are listed below. Any one or combination of these techniques can be used, but they should be carried out in the order shown, as the first two techniques are non-destructive.

1.   The plate can be viewed under an ultraviolet lamp to show any uv-active spots.

2.   The plate can be stained with iodine. This can be achieved rapidly, by shaking the plate in a bottle containing silica and a few crystals of iodine. The iodine will stain any compound that reacts with it and so is especially good for visualizing unsaturated compounds. Most spots show up within a few seconds, but the stain is not usually permanent.

3.   The plate can be treated with one of the reagents listed below and then heated to stain the spots. The reagent can be sprayed onto the plates, but this technique is quite hazardous and it is more effective for them to be dipped in the reagent. To do this, first let the tlc solvent evaporate, then holding the edge of the plate with tweezers, immerse the plate as completely as possible in the stain and remove it quickly. Rest the edge of the plate on a paper towel to absorb the excess stain before heating carefully on a hot plate or with a heat gun, until the spots show. This method is irreversible and so should be done last. When glass plates

are used the spots can sometimes be seen more clearly from the glass side of the plate.

Stains	Use/comments
Vanillin	Good general reagent, gives a range of colours.
PMA	Good general reagent, gives blue-green spots.
Anisaldehyde	Good general reagent, gives a range of colours.
Ceric sulphate	Fairly general, gives a range of colours.
DNP#	Mainly for aldehydes and ketones, gives orange spots.
Permanganate#	Mainly for unsaturated compounds and alcohols, gives yellow spots.

\# - do not usually require heating.

Recipes for visualization reagents.

Vanillin:	vanillin (6g) in ethanol (250ml) + conc. H_2SO_4 (2.5ml).
PMA:	phosphomolybdic acid (12g) in ethanol (250ml).
Anisaldehyde:	anisaldehyde (6g) in ethanol (250ml) + conc. H_2SO_4 (2.5ml).
Ceric sulphate:	15% aqueous H_2SO_4 saturated with ceric sulphate.
DNP:	2,4-dinitrophenolhydrazine (12g) + conc. H_2SO_4 (60ml) + water (80ml) + ethanol (200ml).
Permanganate:	$KMnO_4$ (3g) + K_2CO_3 (20g) + 5% aqueous NaOH (5ml) + water (300ml).

Solvent systems and polarity

The relative distance which a compound travels up a tlc plate depends on two factors, its polarity and that of the solvent. In the same solvent, the more polar the compound, the more tightly it is bound to the silica (or alumina) and the less it travels. There are some common trends, and with experience, it is often possible to predict whether a product will be less or more polar than the starting material. For example when a ketone, or ester is reduced, the resultant alcohol is almost always significantly more polar, and a clean transformation to a lower running spot will indicate a successful reaction. If the polarity of the solvent used for elution is increased the spots will move further up the plate and the distance between the centres of the spots usually increases, up to about half way up the plate. However, the spots also become more diffuse, the further they travel, so an R_f value of about 0.4 is normally the optimum for analytical purposes.

The best tlc solvent system for a particular compound or mixture can only be determined by trial and error. However it is good practice to stick to a 'standard' solvent mixture, which can be used most of the time and which you are familiar with. The most widely used solvent mixtures are based on a non-polar hydrocarbon, such as 40/60 petroleum ether or hexane, with a polar constituent added in a proportion which gives a suitable polarity. Probably the most popular 'universal' tlc system is petroleum ether - ethyl acetate, the polarity of which is easily adjusted by changing the proportions of the two solvents. If the compounds being analyzed will not travel in ethyl acetate mixtures, a more polar solvent such as ethanol is used as the additive. On the other hand, if the compounds travel too far a less polar additive such as petroleum ether is used.

The degree of separation between compounds will also vary according to the solvent used, so if compounds do not separate or give poor separation in one system, different systems should be tried. Where there are a number of compounds in a mixture it may be best to use two or more different systems, for resolving the polar and non-polar components.

Common tlc solvents fall into one of three categories based on polarity, with smaller variations within each category.

Very polar solvent additives: methanol > ethanol > isopropanol

All much more polar than:

Moderately polar additives: acetonitrile > ethyl acetate > chloroform > dichloromethane > diethyl ether > toluene

All much more polar than:

Non-polar solvents: cyclohexane, petroleum-ether, hexane, pentane

Most solvent systems consist of one of the non-polar solvents together with a solvent from one of the other classes. However, for very polar compounds, one of the moderately polar solvents can be used as the less polar constituent. An example of this is chloroform - methanol mixtures, which are useful for highly hydroxylated compounds. The chlorinated hydrocarbons are also commonly used as single components.

Rf values

The R_f value of a compound depends upon the conditions under which the plate was run and is only accurate to about 20%, therefore it is best to compare compounds on the same plate, and run a mixed spot. However, it is useful to record the R_f value of a compound, *remembering to quote the solvent system.*

$$R_f = \frac{\text{Distance of centre of the spot from the baseline}}{\text{Distance of solvent front from baseline}}$$

Multiple elutions

It is sometimes useful to elute a particular tlc plate several times and so improve the separation of closely running spots. In practice this is done by eluting the tlc plate as normal, removing it from the tlc tank and allowing the solvent to evaporate from it and then re-eluting the plate as before. Eluting a plate n-times is effectively the same as running a plate n-times its length.

Running acidic or basic compounds

Acidic and basic compounds often streak up the tlc plate. However, they will usually form distinct spots if, for acids a small amount of a carboxylic acid (e.g. acetic acid) is added to the solvent system, and for bases a small amount of amine (e.g. triethylamine) is added.

Running acid sensitive compounds

The silica on tlc plates is acidic in nature, and so compounds that are sensitive to acid may well decompose on tlc. There are several ways of getting around this problem, you can use alumina tlc plates (these suffer from the disadvantage that resolution is generally not as good, and the plates are basic in nature), or alternatively you can add a small amount of an amine (usually ammonia or triethylamine) to the solvent mixture to neutralize the acidic sites on the silica.

Checking for decomposition

Some organic compounds do decompose on silica to some extent; if you suspect that this is happening you can check by running a two-dimensional plate. This is done by cutting a square plate (ca. 5cm x 5cm) and spotting the compound in the bottom left-hand corner (ca. 0.5cm from the bottom) as shown in Fig. 9.16a. The plate is then eluted as normal to give the spots in a line up the left-hand side of the plate. The plate is then removed from the tlc tank and the solvent allowed to evaporate. It is then placed back in the tank with the line of spots along the bottom, and re-eluted (Fig. 9.16b).

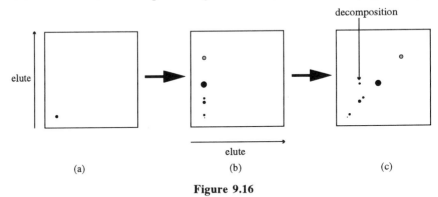

Figure 9.16

The result, if no decomposition is occurring will be the tlc running diagonally across the square plate, however if decomposition is occurring, the spot due to the unstable compound will show decomposition products off-diagonal (Fig. 9.16c).

It is very useful to carry out this test before any form of preparative chromatography, if you suspect that a compound may be labile on silica.

9.3.2 High performance liquid chromatography (hplc)

There is a broad range of hplc techniques, and many different types of equipment. It is beyond the scope of this book to describe in great detail the methods for operating the equipment. This section will therefore focus on some of the ways that hplc can be used to aid the synthetic organic chemist.

Description of hplc

The general arrangement of an hplc system is fairly simple, as shown in Fig. 9.17. Solvent is pumped from a reservoir through a piston pump which controls the flow rate. From the pump the solvent passes through a pulse damper which removes some of the pulsing effect generated in the pump and also acts as a pressure regulator. In between the pulse damper and the column there is an injection valve which allows the sample to be introduced into the solvent stream.

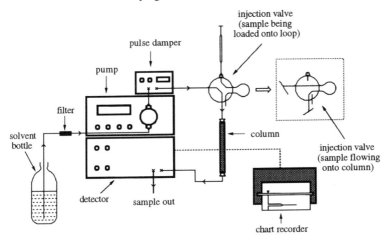

Figure 9.17

In the 'load' mode the solvent by-passes a sample loop, into which the sample is injected from a syringe. On switching to 'inject', the solvent stream is diverted through the load loop, introducing a very accurately measured volume of the sample solution onto the column. Components are separated on the hplc column in exactly the same way as they would be on a tlc plate, the less polar compounds running faster and coming through first. The effluent from the column passes into a detector (usually an ultraviolet or refractive index) which produces a signal on the chart recorder when a component is present.

The time at which the compound comes off the column is characteristic of that particular material, and is referred to as the retention time.

The area under any peak on the chart recorder is proportional to the quantity of that component and the method is therefore quantitative.

Uses of analytical hplc

Finding a tlc system and running a sample can be done very quickly and for this reason tlc is the normal method of choice for routine reaction monitoring. However, there are occasions when it is worth spending the time to set up an hplc system for reaction monitoring, especially if, as in many modern synthetic labs, you have a system close to hand. One reason to use hplc is that the compounds in which you are interested do not separate very well on tlc. The other common reason is that you require a quantitative technique. This may be the case if you are trying to optimize a reaction to maximize the quantity of one product over another, and for this type of extended study it is well worth the time it takes to set up the system. For most synthetic purposes it is the relative, rather than the absolute proportions of substances which are important, and if that is the case a simple comparison of integrated peak areas may be all that is required. If accurate quantification is needed, then a calibration is required and this can be done using an internal standard (see below).

Another common use of hplc is for identification of a compound by comparison with a known substance. Under a specific set of conditions (solvent, flow rate and quantity applied) any compound will have a specific retention time and this can therefore be used as a characteristic of the compound. However, just as a mixed spot should be always be run when comparing substances on tlc, so with hplc a single enhanced peak should be observed when the comparison substance and the unknown are injected as a mixture. Again caution should be used, since a single peak is not absolute proof that compounds are the same.

Preparative hplc is now becoming widely used in organic chemistry for separating compounds with very similar polarity (see Chapter 11 for more details). Before committing all your material to a preparative column it is always best to run a small quantity of the sample on an analytical column, in order to work out the best conditions. Indeed, columns are produced in various sizes which are directly comparable with one another.

If you are monitoring a reaction by hplc and you want to know the identity of one or more of the products, you can often separate a few milligrams from a few runs on the analytical column, which is enough to get a full range of spectral data. On simple hplc systems this can be done manually by collecting the effluent from the column when the peak of interest is coming off, and repeating several times. On more sophisticated systems a fraction collector is often incorporated and in some cases injections can be made automatically, so that the system can be set up to collect a particular peak or peaks over a large number of runs.

Several methods are now available for coupling an hplc system to a mass spectrometer, so that a mass spectrum is produced for each peak thus providing some structural information.

Quantitative analysis

For any type of detector each compound will have a different response, but for a particular compound, the area under its peak is directly proportional to the mass of material which produced the peak. If you are using a uv detector to analyze two compounds which have the same chromophore and extinction coefficients, then you can compare peak areas directly to determine the proportions of each compound.

Since the peak area is proportional to the mass of a compound, it is possible to take a known mass of a compound, make a standard solution and inject specific quantities, to work out the proportionality constant. However, a more accurate method of calibration is to use an internal standard. To do this follow the procedure below:

1. Choose as a standard a readily-available stable compound, with a retention time away from the peaks of interest.

2. Make up at least three mixtures containing known quantities of the standard and each of the compounds which are to be analyzed.

3. Run the mixtures and measure the areas of each peak -
   The mass of material under any peak, y, is:

$$M_y = k_y \times Area_y$$

now, comparing the area under the standard peak with that under the unknown:

$$M_y/M_{(std)} = k_y/k_{(std)} \times Area_y/Area_{(std)}$$

Using this equation we can work out $k_y/k_{(std)}$ which is a constant k_y, known as the correction factor for compound y. Using data from each of the runs, the average correction factor for each compound is calculated.

4. Now if we want to calculate the quantity of a compound in a mixture, the mixture is spiked with a known quantity of the standard and the following equation is used:

$$M_y = k_y/M_{(std)} \times Area_y/Area_{(std)}$$

Peak shape

For good quantitative results from analytical hplc (or gc, see below) you should aim to produce chromatographs with symmetrical peaks. Tailing of the peaks is usually caused by overloading and can thus be avoided by reducing the quantity of sample applied. If this does not solve the problem and the tail of a component is long and drawn out, there may be an incompatibility between the compound and the stationary phase, a problem which is less easy to rectify.

Hplc can also be used for preparative separations and this is described in Chapter 11.

9.3.3 Gas-liquid chromatography (gc, glc, vpc)

Gas-liquid chromatography is becoming increasingly popular for reaction monitoring and for analysis of reaction products. It can be used for the analysis of any compounds which are volatile below about 300°C and thermally stable. It is not the intention of this section to give a detailed description of gc instrumentation, but simply to outline some of the uses of the technique for reaction monitoring and related work.

Description of gas chromatography

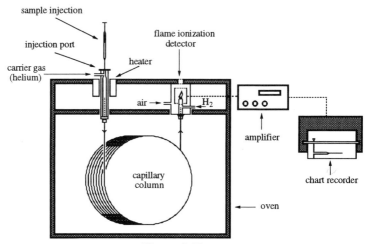

Figure 9.18

Gas chromatography is a very sensitive technique requiring only very small amounts of sample (10^{-6}g). A solution of about 1% is sufficient and a few microlitres of this is injected into a heated injector block. A stream of carrier gas, usually helium, passes through the injector and sweeps the vapours produced onto the column, which is contained in an oven. The temperature of the oven can be accurately controlled and can either be kept constant or increased at a specified rate. Separation of the components in gc is not based on the principle of adsorption, as it is in liquid chromatography, but on partition. A gc column is rather like an extremely effective distillation column with the relative volatility of the components being the main factor which determines how quickly they travel through the column.

The stationary phase of the column is a very high molecular weight, non-volatile oil, which has a very large surface area (see below for more detail about columns) and the gaseous components of the mixture are partitioned between the oil and the carrier gas at different rates. Thus the components are separated along the length of the column and emerge as discrete bands. The gas stream passing out of the column enters a flame ionization detector (FID) which produces an electric current when a compound is burned in the flame. The electric current is amplified to produce a peak on the chart recorder. These detectors are very sensitive and the response produced is proportional to the quantity of material being burned, thus the peak area is proportional to the quantity of sample. As with hplc, the time taken for a particular substance to reach the detector is characteristic of that substance, and referred to as the retention time.

Types of gc column

A gc column must contain a liquid stationary phase with a large surface area, which must be supported, so that it stays in the column and the gas can pass through it. Packed columns are the traditional type, they are made from metal or glass coils, and have an internal diameter of about 2 to 4mm. The stationary liquid phase is coated on particles of solid with which the column is packed. There are a wide variety of stationary phases available ranging from Apiezon greases, which are very non-polar to polyethylene glycols which are very polar. With packed columns the choice of stationary phase is often critical for good separation and a good deal of experimentation is usually required before the best material is found.

The development of modern capillary columns has led to improved resolution and has also simplified the process of running gcs considerably. The columns are normally made from fused silica capillary with an inside diameter of between 0.2 and 0.5mm, and are polymer coated. They have no packing, but instead the liquid stationary phase is bonded to the inside wall of the capillary, and this allows gas to flow very easily. Because of this the columns can be made much longer than packed columns (between 12 and 100m) and they are typically ten times as efficient. Capillary columns give extremely high sensitivity and only a very small quantity of material is required. For this reason the injector normally incorporates a 'splitter', so that only a small portion of the sample injected actually enters the column.

The development of capillary columns has been largely responsible for the increased use of gc for monitoring organic reactions and for product analysis. The increased sensitivity of these columns is one important reason for this, but the fact that they are very simple to set up and operate is perhaps more significant. The type of stationary phase is not so critical as with packed columns and the gas flow rate is essentially determined by the column, so the instrument can be operated successfully with minimal prior expertise. For most purposes relating to preparative organic chemistry it is sufficient to rely on just two types of column, one non-polar (such as a BP1) and one polar column (such as a BP20).

Uses of gc for reaction monitoring and product analysis

Capillary gc instruments are so simple to use that, provided there is one close by, monitoring a reaction by gc is almost as quick as running a tlc. It is common to turn to gc monitoring when tlc does not provide resolution between starting material and product or between one product and another. Gc will usually separate components which co-run on tlc. We also find that some compounds, such as amines, which do not run very well on tlc, can be analyzed very easily by capillary gc.

Gc also provides quantitative analysis and is widely used for determination of product ratios from diastereoselective reactions, down to about 200:1. This makes it an ideal technique for optimization studies, where a large number of small-scale reactions are carried out under different conditions and product ratios are measured simply by syringing out a few microlitres from each and then injecting them into the gc instrument.

For absolute quantitative studies the gc instrument can be calibrated in exactly the same way as described for the hplc instrument. Some people use this technique to measure theoretical yields for reactions, although it is always preferable that isolated yields are quoted.

The identity of a compound can often be determined by gc, if an unknown has the same retention time, and co-runs with a known compound when the two are injected as a mixture, but just as with tlc and hplc co-running, caution should be exercized. A very powerful structure analysis technique is gc mass spectrometry (gc-ms) and when capillary gc is used this is a very simple and quick form of analysis. The mass spectrometer simply acts as the detector, but as well as providing a chromatograph, a mass spectrum of each individual component is obtained. Capillary gc-ms is a very sensitive technique and mass spectroscopic data can be obtained even on very minor components of a mixture.

One example of how gc and gc-ms can be useful is shown in Fig. 9.19. The first time this reaction was carried out, tlc analysis indicated that the starting material had been completely transformed to a single product, which ran as one spot in a number of solvent systems, but the product did not appear to be pure by ^{1}H nmr spectroscopy. When a gc-ms was run on the reaction product, two compounds (which ran together on tlc) were separated and easily distinguished as compounds A and C. Also, 3% of isomeric compound B was identified and although this runs separately from A and C on tlc it had not been detected previously due to lack of sensitivity.

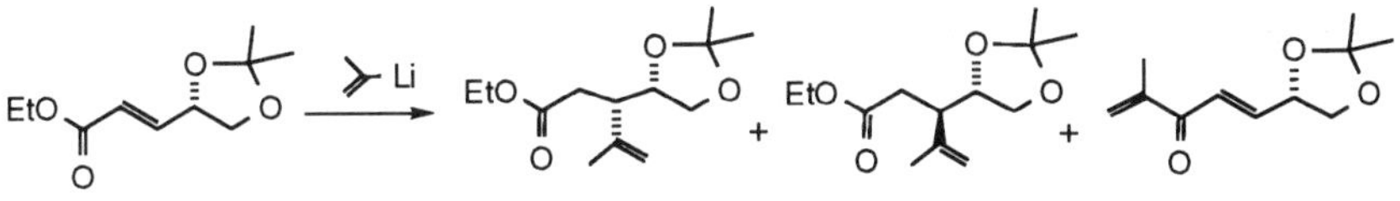

Figure 9.19

Having discovered the identity of the reaction products by gc-ms, product ratios for a series of reactions were obtained using simple gc and we were able to find optimum conditions for generation of compound A.

In this section we have not endeavoured to give a comprehensive review of methods which can be used for reaction monitoring, but we have described the most universal and commonly used modern techniques. Various other monitoring methods can be devised for specific reactions. For instance ultraviolet (uv) spectroscopy can be used if one strong chromophore is being converted to another, but the disadvantage of spectroscopic methods is that they do not indicate how many products are being produced. Nmr spectroscopy can also be employed and this is mentioned further in Chapter 12.

9.4 Reactions at other than room temperature

In many cases it is necessary to carry out reactions at non-ambient temperatures. Usually reactions that are exothermic or involve the intermediacy of thermally-unstable species have to be carried out at low temperatures, typically in the range 0°C to -100°C. Similarly, reactions that are endothermic or have high energies of activation have to be carried out at higher temperatures, typically in the range 30°C to 180°C although in some cases temperatures exceeding 300°C may be necessary. This section deals with the techniques involved in such situations.

9.4.1 Low temperature reactions

Reactions are usually carried out below room temperature by placing the reaction vessel in a cooling bath. In general this is done by placing a cooling mixture into a lagged bath, and then immersing the reaction vessel in the cooling mixture to a depth that ensures the reaction contents are below the level of coolant (Fig. 9.20). The temperature of the coolant can be monitored by means of a low temperature thermometer immersed in the bath.

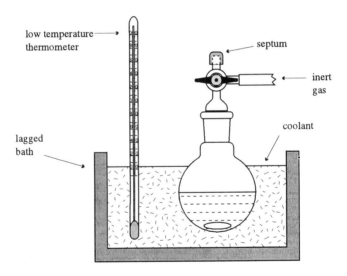

Figure 9.20

It should be noted when carrying out reactions using cooling baths, that the reaction itself may not be at the same temperature as the bath due to exothermic processes taking place, so where possible the internal reaction temperature should also be monitored. A particularly convenient way of doing this is to use a digital low-temperature thermometer. These are commercially available, and most come with a hypodermic probe which can be inserted into the reaction flask through a septum (Fig. 9.21).

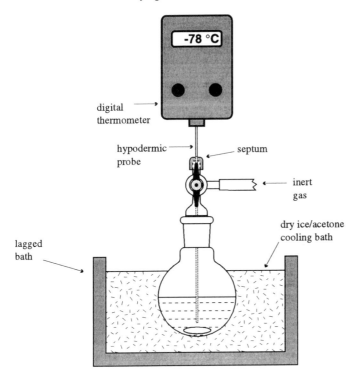

Figure 9.21

The temperature can then be noted and the probe withdrawn. This is often a much easier procedure than the more usual method of setting up the reaction apparatus to include an internal thermometer, and is particularly useful for small-scale set-ups where an internal thermometer cannot be used.

In general all low temperature reactions should be done under inert atmosphere (nitrogen or argon) to avoid atmospheric moisture being condensed into the reaction mixture.

The three main types of cooling mixtures are described below:

i)   Ice-salt baths. Various salts or solvents can be mixed with crushed ice to produce sub-zero temperatures. In practice, temperatures ranging from 0°C to -40°C can be obtained (see Table 9.1); however at the lower temperatures the cooling mixture consists of granular ice-salt particles with little or no liquid, and so this can result in poor thermal contact with any vessel immersed in it. For the lower temperatures, where careful temperature control is important, it is preferable to use a

liquid or slush coolant which has good thermal contact with any vessel
immersed in it.

Table 9.1: Ice-based cold baths[1]

Additive	Ratio (ice/additive)	Temperature °C
Water	1:1	0
NaCl	3:1	-8
Acetone	1:1	-10
$CaCl_2.6H_2O$	4:5	-40

ii)   Dry ice-solvent baths. Solid carbon dioxide (dry ice) is commercially
available as pellets or blocks, and forms very good cooling mixtures
when combined with a variety of organic solvents (see Table 9.2). In
practice the baths are prepared by adding the dry ice pellets carefully to
a bath containing the requisite solvent until the temperature required is
reached. The temperatures quoted in Table 9.2 refer to baths in which
an excess of dry ice is contained in the solvent. In this case cooling
mixtures ranging from -15°C to -78°C can be achieved.

Table 9.2: Dry-ice cold baths[2]

Solvent	Temperature °C	Solvent	Temperature °C
Ethylene glycol	-15	Chloroform	-61
Carbon tetrachloride	-25	Ethanol	-72
Heptan-3-one	-38	Acetone	-78
Acetonitrile	-42		

iii)   Liquid nitrogen slush baths. Slush baths are made by adding liquid
nitrogen carefully to a solvent contained in the bath, with continuous
stirring (glass rod, not a thermometer). The coolant should become the
consistency of ice-cream, and stirring should prevent any solidification.
Again a variety of different liquids can be used to give temperatures
ranging from 13°C to -196°C (see Table 9.3).

Such cooling systems can often be left for several hours if the cooling
bath is well lagged, however for longer periods (overnight) some form of
mechanical cooling is usually necessary. In such instances, the reaction

1.   A.J. Gordon and R.A. Ford,   "The Chemist's Companion", J. Wiley and Sons,
     New York, 1972.
2.   A.M. Phillips and D.N. Hume, *J. Chem. Ed.*, 1968, **54**, 664.

vessel can be placed in a refrigerator, or cooled by the use of a portable commercial refrigeration unit.

Table 9.3: Liquid nitrogen slush baths[3]

Solvent	Temperature °C	Solvent	Temperature °C
p-Xylene	13	Chloroform	-63
p-Dioxane	12	Isopropyl acetate	-73
Cyclohexane	6	Butyl acetate	-77
Formamide	2	Ethyl acetate	-84
Aniline	-6	2-Butanone	-86
Diethylene glycol	-10	Isopropanol	-89
Cycloheptane	-12	n-Propyl acetate	-92
Benzyl alcohol	-15	Hexane	-94
o-Dichlorobenzene	-18	Toluene	-95
Carbon tetrachloride	-23	Methanol	-98
o-Xylene	-29	Cyclohexene	-104
m-Toluidine	-32	Isooctane	-107
Thiophene	-38	Carbon disulphide	-110
Acetonitrile	-41	Ethanol	-116
Chlorobenzene	-45	Methyl cyclohexane	-126
m-Xylene	-47	n-Pentane	-131
Benzyl acetate	-52	Isopentane	-160
n-Octane	-56	Liquid nitrogen	-196

9.4.2 Reactions above room temperature

1. Reactions in a sealed tube

Reactions above room temperature usually require modifications to the standard equipment set-up. In some instances the reaction can be performed in a sealed tube, usually made of thick-walled glass. The reaction mixture is placed in the tube which is then sealed, placed in an oven and heated to the appropriate temperature (Fig. 9.22). After the reaction is complete, the tube is cooled, opened and the contents removed. Such a technique is employed when temperatures in excess of the solvent boiling point are required, or for reactions involving extremely volatile compounds.

3.   R.E. Rondeau, *J. Chem. Eng. Data*, 1966, **11**, 124.

This technique naturally requires a high degree of skill, since heating leads to a pressure build-up inside the tube which can result in explosion if there are any flaws in the seal.

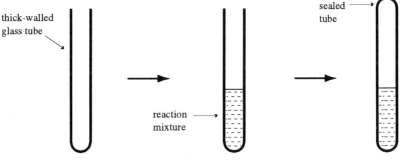

thick-walled glass tube

sealed tube

reaction mixture

Figure 9.22

An alternative to the all-glass sealed tube is to use a reaction tube, which consists of a thick-walled glass tube with a Teflon screw seal at the top (Fig. 9.23). This serves as a re-usable sealed tube apparatus, and commercial versions of this apparatus are available. It also features a useful side-arm that allows evacuation or purging with an inert gas prior to sealing the tube. It should be noted that some apparatus of this type use an O-ring seal. In such instances it is essential to make sure that the O-ring is made of a material inert towards the reaction contents otherwise the seal may fail.

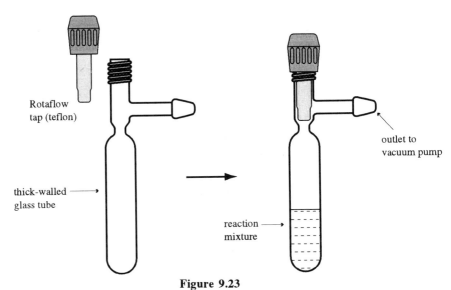

Rotaflow tap (teflon)

thick-walled glass tube

reaction mixture

outlet to vacuum pump

Figure 9.23

2. Reactions involving the use of a condenser

For most reactions above room temperature, an open system which does not lead to a build-up of pressure is employed. This usually consists of a reaction vessel protected with a condenser (Fig. 9.24). The condenser is used to prevent the evaporation of volatile materials (usually the solvent) from the reaction mixture.

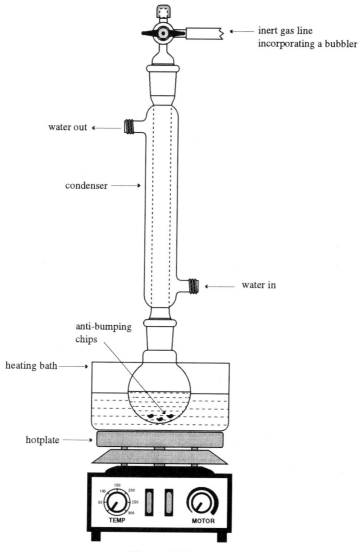

Figure 9.24

There are many different designs of condenser available, and the type
used depends upon the nature of the reaction involved. The most common
designs of condenser are the Liebig condenser (Fig. 9.25a), the coil
condenser (Fig. 9.25b), the double-jacketed coil condenser (Fig. 9.25c), and
the cold-finger condenser (Fig. 9.25d). Other condensers available tend to
be simple modifications of these three types.

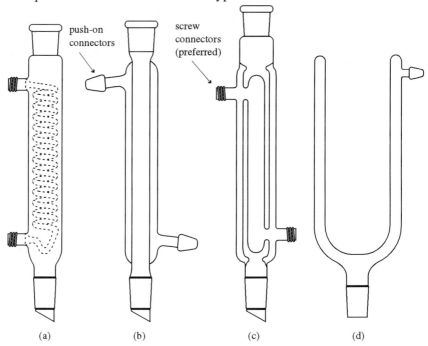

Figure 9.25

The Liebig condenser, the coil condenser, and the double-jacketed coil
condenser are similar in design and function. They are water-cooled *via*
connection to a cold-water tap, in the case of the Liebig condenser the water
flows in at the bottom and flows out at the top giving a jacket of cold water
around the condenser stem and leading to a cold surface on the inside. Any
volatile materials in the reaction condense on the cold outer surface and run
back into the reaction mixture. The coil condenser functions in a similar
way except that the cold surface is now on the inside of the condenser. This
can offer an advantage in particularly humid locations because there is less
tendency for atmospheric moisture to condense on the outside of the
condenser and run down over the reaction vessel. The double-jacketed coil

condenser is also water-cooled, again water flows in at the bottom and out at the top. This condenser design tends to be more efficient than the other two because it provides a greater area of cold surface. Consequently it is preferred when low boiling materials (≤ 40°C) are involved.

The cold-finger condenser is rather different from the above three. It is cooled by either solid carbon dioxide / acetone (-78°C) or liquid nitrogen (-196°C). The coolant is placed in the top of the condenser and more coolant is added as required. This results in an extremely cold surface on the inside of the condenser. Condensers of this type are usually employed for reactions that involve solvents or components that boil at or below room temperature (e.g. liquid ammonia, b.p. -33°C), although they can be used for higher boiling materials as well.

3. Heating devices

As with low-temperature reactions, a common method of increasing the temperature of a reaction is to place the reaction flask in a bath (Fig. 9.23). In this case the contents of the bath are then heated, usually using a hotplate.

There are five commonly employed heating baths: water, silicone oil, Woods metal, flaked graphite, and sand. Water baths consist of a Pyrex glass container filled with water, and are used for reactions requiring temperatures up to 100°C. Silicone oil baths are similar in design but can be used for temperatures up to about 180°C, although the maximum temperature in this case depends upon the precise type of oil employed. Woods metal is a commercially available alloy (50% Bi, 25% Pb, 12.5% Sn, 12.5% Cd) that melts at 70°C. It has excellent thermal properties and so is a safe material for use at quite high temperatures (up to 300°C). It is normally used in a steel container, and again heated via a hot-plate device. Similarly, flaked graphite and sand can be used in heating baths up to 300°C, although they cool slowly compared with the Woods metal.

Reactions can also be heated by an electric heating mantle (Fig. 9.26), although in general this is less satisfactory since it is more difficult to control the temperature of the mantle surface and excessive heating of the reaction flask can result. Because of this, where possible heating mantles should be avoided for reactions below 100°C although they can be used if necessary.

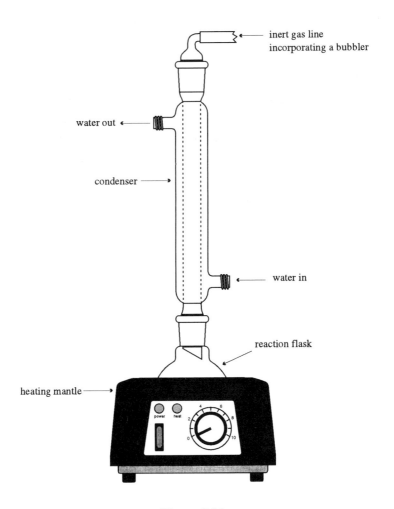

Figure 9.26

9.5 Driving equilibria

A number of important organic reactions involve equilibria and do not give
good yields unless the equilibrium can be shifted to the product side. An
equilibrium can be driven to the right by using an excess of one of the
reactants, by continuously removing one of the products, or by changing the
temperature or the pressure at which the reaction is carried out. In most
cases use of excess reagent or removal of a product can be achieved using
normal apparatus and techniques. This section is concerned with two

methods which involve special apparatus: Dean and Stark traps, and high pressure reactors.

9.5.1 Dean-Stark traps

Perhaps the most commonly encountered equilibrium reactions are those involving water as a reactant or product. Driving such equilibria by using excess water (e.g. hydrolysis reactions) is easy, but driving equilibria by removing water (e.g. in ester or acetal formation) can be more difficult. An excellent device for the continuous removal of water from a reaction mixture is the Dean-Stark trap (Fig. 9.27).

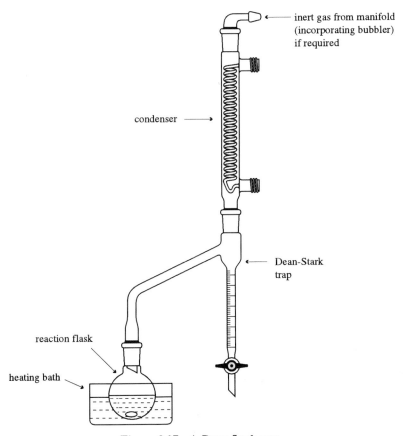

Figure 9.27   A Dean-Stark trap

The apparatus is assembled as shown and the reaction is conducted in a solvent which forms an azeotrope with water (almost invariably a hydrocarbon such as toluene). When the mixture is heated the solvent/water azeotrope distils over and, on condensing, is collected in the trap. The water then separates and the light organic solvent flows back into the reaction flask. It is usually easy to monitor the progress of the reaction either by recording the volume of water produced or by waiting until the characteristically milky heterogeneous azeotrope is no longer produced.

A Dean-Stark trap can also be used to remove volatile alcohols, such as methanol and ethanol, which *are* miscible with many organic solvents. They can nevertheless be removed by placing 5Å molecular sieves (Section 4.3) in the trap, in order to absorb the alcohol. An alternative is to use a Soxhlet extractor containing molecular sieves.

On a small scale, simply placing some activated sieves in the reaction flask is a convenient means of removing water. This method is effective in driving equilibria and is also used to promote reactions which are adversely affected by water.

9.5.2 *High pressure reactions*

A more esoteric technique is to carry out the reaction at very high pressures (>10kbar), thus shifting the equilibrium to the side of the components which have the smaller volume. Some transformation of the products must be carried out as soon as they are removed from the reactor, in order to prevent re-equilibration. A recent review[4] gives an account of some applications of high pressure techniques.

9.6 Agitation

For homogeneous reaction systems at constant temperature, agitation is not normally necessary. However, most reactions involve heterogeneous reaction mixtures and so require some form of agitation to ensure efficient mixing of the reactants. The most commonly used methods of agitation are outlined in this section.

9.6.1 *Magnetic stirring*

Magnetic stirrer machines are commonly available and come in two general types: either a simple stirring machine, or one that also incorporates a

---

4. N.S. Isaacs and A.V. George, *Chem. Br.*, 1987, **23**, 47.

hotplate for heating reaction systems (Fig. 9.28). They consist of a box containing a motor that drives an electromagnet which spins horizontally. Most magnetic stirrers have a flat top to allow cooling or heating baths (see Section 4.6) to be placed on top. Agitation of the reaction mixture is achieved by placing a magnetic stirrer bar (or follower) in the reaction mixture. The reaction vessel is then clamped over the top of the stirrer machine in such a position as to allow the mixture to be stirred by magnetic interaction of the follower with the electromagnet in the stirrer machine. Usually the rate of stirring can be adjusted using a control on the stirrer machine.

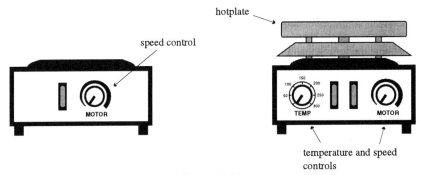

Figure 9.28

The follower consists of a magnet coated with an inert polymer, usually Teflon or PVC, and comes in a variety of shapes and sizes (Fig. 9.29). It is important that the polymer coating does not react with any components in the reaction mixture, and because of this Teflon is the recommended coating for most reaction systems. One notable exception to this guideline are reactions involving metal-ammonia solutions which attack Teflon. The size of the follower is also important, it should be large enough to stir the reaction mixture effectively but not so large that it will not sit flat in the bottom of the reaction flask.

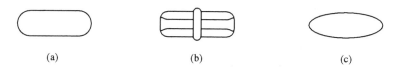

(a)   (b)   (c)

Figure 9.29 Magnetic followers: (a) bar; (b) octagonal; (c) egg-shaped

Because the follower is driven by a magnetic field and has no mechanical connection with the stirrer machine, this method of agitation does not require special modification of a reaction apparatus.

Magnetic stirrer machines are probably the most commonly employed method of agitation for organic reactions, but can become ineffective if particularly viscous systems are encountered. They may also be ineffective if the reaction vessel has to be placed inside another piece of apparatus such as a heating mantle or large cooling bath. In such cases the extra apparatus can effectively shield the reaction flask from the magnetic field created by the stirrer machine. Magnetic stirring can also be a problem for large scale reactions (reaction volumes over one litre), and in such cases mechanical stirring is a valuable alternative.

9.6.2 Mechanical stirrers

Mechanical stirring machines consist of an electric motor clamped above the reaction vessel, that rotates a vertical rod (usually glass, although it can be steel or Teflon). A vane or paddle is attached to the bottom of this rod, and it is this that is responsible for agitation of the reaction mixture (Fig. 9.30) The rod and vane are usually detachable enabling different length rods and different sized vanes to be used as appropriate. As with magnetic stirrer machines, the rate of stirring can usually be adjusted by means of a control on the motor. There are many different designs for the vane, the most common being a crescent shaped piece of Teflon about 5mm thick. This has a slot in it that allows easy attachment to the glass rod (Fig. 9.31). In this design the vane can be rotated about a horizontal axis and so can easily be put through the narrow neck of a round-bottomed flask, then rotated into a horizontal position ready for use.

Because the mechanical stirrer requires a physical attachment to the reaction flask, precautions have to be taken if the reaction is to be carried out under anhydrous conditions, or under an inert gas atmosphere. The usual way of doing this is to use a stirrer guide that allows the rod to enter the reaction flask, but prevents atmospheric gases from doing so. These are constructed from Teflon (Fig.9.32a), and provide a tight fit between a ground glass joint and the glass rod. The tight seal around the rod is achieved by means of an O-ring seal inside the stirrer guide, and a set-up of this type should be good enough to withstand a vacuum of 0.5-0.1 mmHg

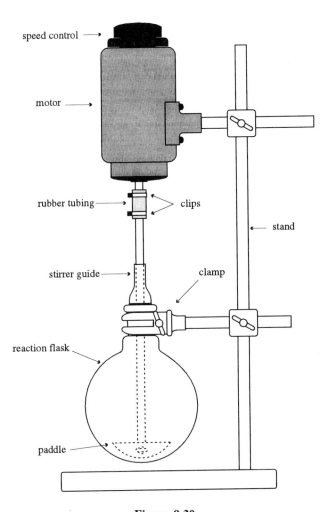

speed control

motor

rubber tubing

clips

stand

stirrer guide

clamp

reaction flask

paddle

Figure 9.30

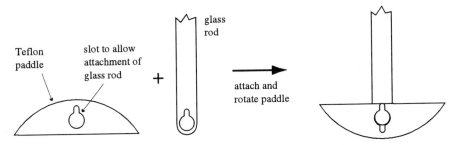

Teflon
paddle

slot to allow
attachment of
glass rod

glass
rod

+

attach and
rotate paddle

Figure 9.31

inside the reaction flask without leaking. Because the guide is constructed entirely from Teflon it does not require lubrication.

Such guides are necessarily expensive, and for many uses that do not require rigorously controlled reaction conditions a glass guide can be employed (Fig. 9.32b). This consists of a precision-ground glass tube attached to a normal ground glass joint. In this case it is essential to use a ground glass rod that forms a good fit with the tube. Oil lubrication is required to allow the rod to rotate, and so this set-up is not suitable for reactions above room temperature that involve volatile solvents which can leach the lubricant into the reaction mixture.

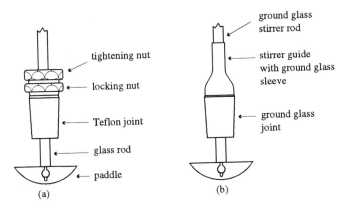

Figure 9.32

Because mechanical stirrers use an electric motor mounted above the reaction flask, there is a serious danger of sparks from the motor igniting volatile flammable solvents. In such instances the use of mechanical stirrers powered by compressed air is recommended.

9.6.3 Mechanical shakers

Mechanical shakers come in many designs, and are simply motors that will shake an attached reaction flask. The flask is usually clamped to the shaker (Fig. 9.33), often with a counter-weight to balance the machine. This is a useful device for reactions that involve vigorous mixing of two immiscible liquids for prolonged periods of time, and can also be employed when efficient mixing of a gas and liquid are required (e.g. hydrogenation); however, for most applications other methods of agitation are preferable.

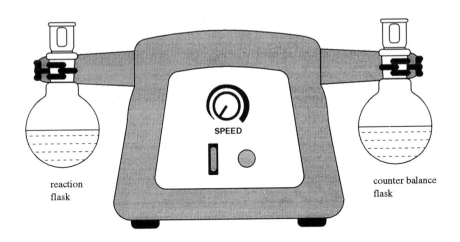

Figure 9.33

9.6.4 Sonication

Ultrasonic waves can also be employed as a means of agitation. The most common arrangement is to use a simple ultrasonic bath, in which the

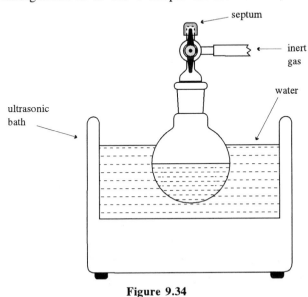

Figure 9.34

reaction vessel is placed (Fig. 9.34), although ultrasonic probes can also be used and are often placed inside the reaction vessel (Fig. 9.35). The latter arrangement is particularly desirable if precise control of the ultrasound frequency is required, or if external control of the reaction temperature is necessary. In both cases ultrasonic waves are generated inside the reaction vessel, causing agitation of its contents. This technique is particularly useful for reactions involving insoluble solids. The ultrasonic waves break up the solids into very small particles facilitating solvolysis and reaction.

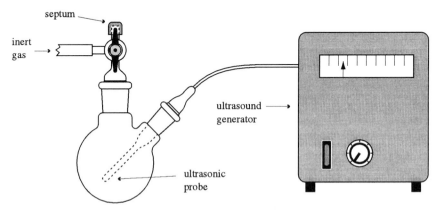

Figure 9.35

CHAPTER 10

Working Up The Reaction

10.1 Introduction

It is important to give some thought to the work up of the reaction before you attempt it. There are several aspects needing to be considered which will be dealt with here. First of all do make sure that the reaction has indeed finished (by careful analysis using your chosen monitoring system). When using tlc analysis it is sometimes difficult to judge by spotting the reaction mixture directly on to the tlc plate. In these cases it is often possible to get a more accurate assessment by withdrawing a small aliquot of reaction mixture by syringe and adding it to a small vial containing a few drops each of diethyl ether and aqueous ammonium chloride. Agitation followed by tlc of the organic phase will often give clean, reliable tlc information. This technique can also be used to screen alternative work up conditions, for example adding to water or aqueous base rather than ammonium chloride solution, or using other organic solvents in place of diethyl ether.

Having satisfied yourself that the reaction has run to completion, or that it is time to end the experiment, the appropriate "quench" is added to the reaction mixture. Choice of this reagent can be very important in determining the yield of desired product, and it is obviously vital to use a reagent or procedure which is safe. Given that the product is expected to be reasonably stable, which usually is the case, then the choice of procedure for quenching the reaction is determined by the reagent(s) used in the reaction. Clearly we cannot cover all possibilities in this Chapter, but general procedures which should cover most of the situations which are likely to be encountered are provided below. The classification is made on the basis of the nature of the reaction mixture which is to be worked up.

10.2 Quenching the reaction

10.2.1 General comments

If the reaction has been carried out under an inert atmosphere then it is advisable to add the quench before exposing the reaction mixture to the air. It is best added as you would add a reagent (dropwise by syringe). If the reaction was run at low temperature, then add the quench at this temperature and allow to warm to room temperature slowly before opening to the air and proceeding with the isolation of the product. If the reaction was run at elevated temperature, allow to cool before adding the quench (still under the inert atmosphere if used). If an exotherm is possible, ensure that the reaction is cooled in a cooling bath (e.g. ice/water) if it is not already in a low temperature bath, and that the quench is added very carefully (if the scale of the reaction is such that an internal thermometer can be used, then do so and keep a close eye on the internal temperature while quenching the reaction).

These general comments apply to the following specific types of reactions. Obviously if a known, fully described literature procedure is being used, then follow the work up and isolation process exactly. Only modify such a procedure if you encounter problems.

10.2.2 Strongly basic non-aqueous reactions

This is a common type of reaction, typical examples would include alkylation using strong bases [e.g. BuLi, (i-Pr)$_2$NLi], many organometallic reagents (e.g. MeLi, Grignard reagents), hydride reducing agents (e.g. LiAlH$_4$, Na(EtOCH$_2$CH$_2$O)$_2$AlH$_2$), and cuprate reactions. The most commonly used quench for this type of reaction is an excess of aqueous ammonium chloride, added slowly, to protonate anions present and to destroy unreacted reagent. This can be added at low temperatures, but if you are concerned about the aqueous solution freezing out (there is usually no need to worry about this in our experience) then use acetic acid as the quench . Make sure that enough is added to destroy all the reagent and to protonate any anions which might be present, but do not add too much as it can sometimes make isolation and purification of the product more difficult.

In the case of aluminium based reagents such as LiAlH$_4$ and (i-Bu)$_2$AlH, alternative procedures are often used to attempt to avoid very fine precipitates which are difficult to filter and which can lead to emulsions. One simple method which often gives good results is to add a saturated aqueous solution of sodium sulphate dropwise with stirring (and cooling!), until a very heavy precipitate is formed (do not add excess solution just to "make sure"). The supernatant can then be decanted and the solid extracted a few times with the reaction solvent, combining all the organics. If you try this, do not dispose of the solid until you are certain that you have obtained a good recovery of material. It might be necessary to extract further to obtain all of the product.

An alternative work-up procedure specifically for LiAlH$_4$ is as follows. The stirred reaction mixture is treated dropwise (*Caution!*), in sequence with the following: a) Water (1ml per g of LiAlH$_4$ used), b) Aqueous sodium hydroxide (15% solution, 1ml per g of LiAlH$_4$ used), and c) Water (3ml per g of LiAlH$_4$ used). This will often produce a granular precipitate which is easy to filter and to wash.

10.2.3 Neutral non-aqueous reactions

"Neutral" is taken to cover all reactions which involve neither strong base nor acid. Such a "neutral" reaction might be in fact slightly acidic (e.g. acid-catalysed ketalisation) or slightly basic (e.g. tosylation of an alcohol using pyridine or triethylamine as the base). Clearly a great many reactions fall into this category and in general the quench can be ammonium chloride solution or water for mildly basic reactions, and dilute sodium hydrogen carbonate for mildly acidic reactions.

If a fairly reactive reagent has been used then add the quench carefully with cooling. For example, if p-toluenesulphonyl chloride has been used to prepare a tosylate it is useful to add a relatively small volume of water and stir for an hour or so to destroy any remaining sulphonyl chloride (be certain to add at least enough to destroy all of the sulphonyl chloride which was used). This will convert the chloride to the sulphonic acid (present as a salt with the amine used in the reaction) which is easy to remove in the subsequent extraction and purification.

As ever, it is important to be certain that your product is unlikely to be sensitive to (or destroyed by) the quench. For example, any reaction which has been driven to completion by removal of water needs some thought. A common example of this is acid-catalysed ketalisation. The product ketal is likely to be sensitive to aqueous acid, so be sure to quench with plenty of aqueous sodium hydrogen carbonate solution before proceeding with the isolation.

10.2.4 Strongly acidic non-aqueous reactions

This type of reaction will usually involve the use of a strong Lewis acid such as $TiCl_4$ or $BF_3.OEt_2$. It is important to be aware that the addition of water to these reagents will be exothermic and usually liberates strong protic acid. If the product is likely to be unaffected by the liberated acid, then water can be used as a quench. It is often the case that the product will be unstable towards this acid, or at least to contain functionality which will be destroyed by acid, and in this case aqueous sodium bicarbonate or carbonate can be used. If a non-aqueous quench is desired then a solution of gaseous ammonia in the reaction solvent can be used. Quench these reactions carefully, with cooling, and be aware that the metals used (Ti and Al particularly) can give rise to insoluble precipitates and be prepared to deal with the problems which these might cause (see below).

10.2.5 Acidic or basic aqueous reactions

To quench these reactions it is usually enough to neutralize with dilute acid or base, but do consider the product which you wish to isolate. For example if you are hydrolyzing an ester then no quench will be required, the desired product will be isolated by the appropriate extraction procedure.

10.2.6 Liquid ammonia reactions

A number of synthetically useful reactions use this as solvent, and usually involve either the use of, or generation of strongly basic species. The usual quench for this type of reaction is to add (*carefully*) an excess solid ammonium chloride, and then allow the ammonia to evaporate (*fume cupboard*).

Conclusion

In conclusion, the correct quench for a reaction is usually simple to determine, given that adequate consideration is given to the reagents used and the product expected. If you are at all unsure, then remove a small quantity of the reaction mixture and add it to a vial containing a few drops of both the quench and the reaction solvent and observe . Tlc analysis (or whatever analytical technique was used to monitor the reaction) will be useful in ensuring that all is well (or otherwise). It is better to "waste" a little product in this way than to lose it all by incorrect choice of quench.

With the reaction successfully quenched, then the isolation of the crude product can be carried out, as outlined below.

10.3 Isolation of the crude product

10.3.1 General comments

The usual isolation procedure involves partitioning the reaction mixture between an organic solvent and water (or aqueous solutions). Before carrying this out it is often useful to remove any solid which is present. It can be particularly important to remove very fine particles, even if they are present in a small amount, as these can lead to emulsions forming in subsequent extractions. This is best achieved by dilution of the (quenched) reaction mixture with the reaction solvent followed by vacuum filtration through a Celite pad, made by packing Celite onto a sintered glass funnel (with the vacuum on). Do wash the pad thoroughly with water, followed by the solvent which you are using since Celite can contain soluble impurities. Simply filter the reaction mixture through this pad, and rinse through two or three times to ensure complete transfer of your product.

If your solvent has reasonable solubility in water then it is advisable to remove most or all of it at this stage on a rotary evaporator. This is not always necessary but omitting it can lead to losses of product to the aqueous phase during extraction. The most commonly encountered examples of this type of solvent are alcohols and tetrahydrofuran, and the problem gets worse as the polarity of the product increases.

The reaction mixture can now be partitioned between water (or an aqueous solution) and an organic solvent. Throughout the extraction procedure do not discard any extract until you have made sure that you have a good mass recovery (on weighing the crude product).

If a low recovery is obtained the rest must be somewhere and it is likely that the product will be in one of the extracts. For most reactions dichloromethane can be used or diethyl ether if a solvent less dense than water is required. If a low recovery is obtained with dichloromethane or diethyl ether, then simply extract the aqueous layers (which you have kept!) with either chloroform or ethyl acetate. These are more powerful solvents and will often extract polar compounds from aqueous solutions. If this fails, try saturating the aqueous extracts with sodium chloride and extracting again with chloroform or ethyl acetate. For especially difficult cases extraction of the aqueous (after saturation with sodium chloride) with a mixture of chloroform and ethanol (2:1) is often successful. Whenever you extract with an organic solvent use several portions rather than one large one as this is much more efficient.

If an amine (for example, triethylamine or pyridine) has been used in the reaction then this can be removed by washing with dilute acid. If the product might be unstable to this, then several extractions with aqueous copper sulphate will often remove the amine.

Occasionally you will encounter the problem of emulsions during extraction. There is no universal solution to this, but a common cause is the presence of very fine particles, which can be removed by filtration through Celite as described above. If this fails, try adding sodium chloride to the aqueous, and try adding another organic solvent.

10.3.2 Very polar aprotic solvents

The most commonly encountered of these are dimethyl sulphoxide (DMSO) and N,N-dimethylformamide (DMF). With these solvents it is often possible to remove most if not all by adding the quenched reaction mixture to a relatively large volume of water, and extracting this several times with ether. The combined organics are then washed with more water. Any remaining DMF or DMSO will need to be removed in further purification of the crude product.

Some reactions use hexamethylphosphoramide (HMPA) as an additive, which can be difficult to remove. HMPA can be extracted from an organic solution by washing with aqueous lithium chloride or bromide provided that the solvent for the organic solution is not chlorinated.

With the extractions completed the organic extract needs to be dried. A lot of the dissolved water can be removed by extracting with saturated brine and it is advisable to do this. The organic extract is then dried fully by addition of either anhydrous magnesium or sodium sulphate and standing until dry, followed by vacuum filtration through a sintered glass funnel.

The solvent is now usually removed using a rotary evaporator, with the final traces being removed using a high vacuum line. Beware when using the high vacuum line, if your product is volatile it will "disappear" into the traps!

At this stage it is very useful (*after weighing!*) to examine the crude product by ir, nmr, and tlc before proceeding with purification as this will give you an idea as to the state of purity amongst other things. The next chapter is given over to methods for purification of the crude product.

CHAPTER 11

Purification

11.1 Introduction

When a product has been isolated from a reaction the next step is to purify it. The degree of purity required will depend on the use for which the sample is intended, a synthetic intermediate might only require rough purification, whereas a product for elemental analysis would require rigorous purification. This section describes the most important purification techniques, crystallization, distillation, sublimation, and chromatography. It is assumed that the reader is familiar with the basic principles of these methods, so the emphasis is on more demanding applications such as the purification of air-sensitive materials, and purifications on a micro-scale.

11.2 Crystallization

11.2.1 Simple crystallization

Simple crystallization of an impure solid is a routine operation, which nevertheless requires care and good judgement if good results are to be obtained. The basic procedure can be broken down into six steps, which are listed below, together with some tips on how to overcome common problems.

1. *Select a suitable solvent*

   Find a suitable solvent by carrying out small scale tests. Remember that 'like dissolves like'. The most commonly used solvents in order of increasing polarity are petroleum ether, toluene, chloroform, acetone, ethyl acetate, ethanol, and water. Chloroform and dichloromethane are rarely useful on their own because they are good solvents for the great majority of organic compounds. It is preferable to use a solvent with a boiling point in excess of 60°C, but the b.p. should be at least 10°C lower than the m.p. of the compound to be crystallized, in order to prevent the solute from 'oiling out' of solution. In many cases a mixed solvent must be used, and combinations of toluene, chloroform, or ethyl acetate, with the petroleum ether fraction of similar boiling point are particularly useful. Consult Appendix 1 for boiling points, polarity (dielectric constant), and toxicity of common solvents.

2. *Dissolve the compound in the minimum volume of hot solvent*

   Remember that most organic solvents are extremely flammable and that many produce very toxic vapour.

   Place the crude compound (always keep a few 'seed' crystals) in a conical flask fitted with a reflux condenser, add boiling chips and a small portion of solvent, and heat in a water bath. Continue to add portions of solvent at intervals until all of the crude has dissolved in the hot/refluxing solvent. If you are using a mixed solvent, dissolve the crude in a small volume of the good solvent, heat to reflux, add the poor solvent in portions until the compound just begins to precipitate (cloudiness), add a few drops of the good solvent to re-dissolve the compound, and allow to cool. When adding the solvent it is very easy to be misled into adding far too much if the crude is contaminated with an insoluble material such as silica or magnesium sulphate.

3. *Filter the hot solution to remove insoluble impurities*

   This step is often problematic and should NOT be carried out unless an unacceptable (use your judgement) amount of insoluble material is suspended in the solution. The difficulty here is that the compound tends to crystallize during the filtration so an excess of solvent (*ca.* 5%) should be added, and the apparatus used for the filtration should be preheated to about the boiling point of the solvent. Use a clean sintered funnel of porosity 2 or 3, or a Hirsch or Büchner funnel, and use the minimum suction needed to draw the solution rapidly through the funnel. If the solution is very dark and/or contains small amounts of tarry impurities, allow it to cool for a few moments, add *ca.* 2% by weight of decolorising charcoal, reflux for a few minutes, and filter off the charcoal. Charcoal is very finely divided so it is essential to put a

1cm layer of a filter aid such as Celite on the funnel before filtering the suspension.   Observe the usual precautions for preventing crystallization in the funnel.   Very dark or tarry products should be chromatographed through a short (2-3cm) plug of silica before attempted crystallization.

4.   *Allow the solution to cool and the crystals to form*

This is usually straightforward except when the material is very impure or has a low m.p. (< 40°C) in which case it sometimes precipitates as an oil.   If an oil forms it is best to reheat the solution and then to allow it to cool slowly.   Try scratching the flask with a glass rod or adding a few 'seed' crystals to induce crystallization, and if this fails try adding some more solvent so that precipitation occurs at a lower temperature. If nothing at all precipitates from the solution, try scratching with a glass rod, seeding, or cooling the solution in ice-water.   If all these fail, stopper the flask and set it aside for a few days, patience is sometimes the best policy.

5.   *Filter off and dry the crystals*

When crystallization appears to be complete filter off the crystals using an appropriately sized sintered glass funnel.   It is very important to wash the crystals carefully.   As soon as all of the mother liquor has drained through the funnel, remove the suction and pour some *cold* solvent over the crystals, stirring them if necessary, in order to ensure that they are thoroughly washed.   Drain off the washing under suction and repeat once or twice more.   After careful washing allow the crystals to dry briefly in the air and then remove the last traces of solvent under vacuum, in a vacuum oven, in a drying pistol, or on a vacuum line. Take care to protect your crystals against accidental spillage or contamination.   If they are placed in a dish, a beaker, or a sample vial, cover with aluminium foil, secure with wire or an elastic band, and punch a few small holes in the foil.   If the crystals are in a flask connected directly to a vacuum line, use a tubing adapter with a tap and put a plug of glass wool in the upper neck of the adapter so that the crystals are not blown about or contaminated with rubbish from the tubing, when the air/inert gas is allowed in.   If a relatively high boiling solvent such as toluene was used for the crystallization it is essential to heat the sample under vacuum for several hours to ensure that all of the solvent is removed.

11.2.2 Small scale crystallization

When small samples (less than 200mg) are to be crystallized it is particularly important to remove traces of insoluble material and to avoid contamination by dust, filter paper, etc. Thus it is essential to use carefully cleaned glassware and purified solvents, and to filter the hot solution. The apparatus shown in Fig. 11.1 is convenient for small scale (5-200mg) crystallizations. Place the sample in the bulb and wash it with a few drops of solvent. The procedure is as follows:

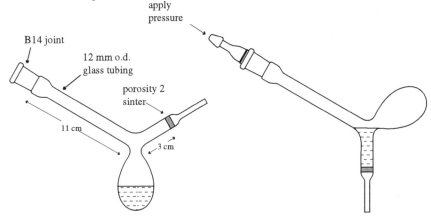

Figure 11.1

1. Heat the bulb in an oil bath or a water bath so that the solvent refluxes up to the level of the sintered disc. Add more solvent, in small portions, until the compound has dissolved.
2. Remove the apparatus from the bath, wipe the bulb to remove oil or water, and quickly filter the hot solution into a clean receiver by pressurizing the vessel using hand bellows or an inert gas line. Filtering under pressure in this way avoids the problem of unwanted crystallization, and reduces transfer losses. The hot solution can be filtered into a small conical flask, but on a scale of 100mg or less, a Craig tube (Fig. 11.2) gives better recovery because it allows the crystals to be recovered without another filtration.

3. Filter the hot solution into a suitably sized Craig tube and cover the tube with aluminium foil while crystallization occurs (Fig 11.2a).

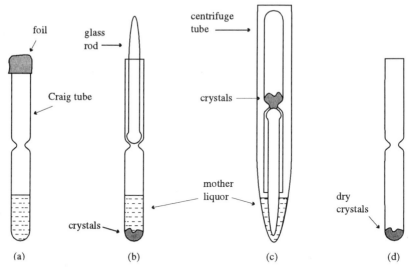

Figure 11.2

4. When crystallization is complete fit the matching glass rod (a close fit is essential) into the Craig tube and secure it tightly with a rubber band (Fig. 11.2b). Place the inverted assembly in a centrifuge tube and centrifuge for a few minutes (remember to use a counter-balancing tube and solvent).

5. After centrifuging, the mother liquor will have been forced into the centrifuge tube (Fig. 11.2c). Remove the Craig tube from the centrifuge tube, remove the glass rod, then tap the tube to so that the crystals fall to the bottom (Fig. 11.2d). Cover the tube with foil, and dry the crystals under vacuum.

6. If necessary the product can be recrystallized again in the same tube by repeating the procedure above. The crystals can be dissolved in the Craig tube, in the minimum volume of hot solvent as usual, and the neck of the tube will act as a condenser. Care is required though, and the tube should be no more than one third full of solvent in order to avoid losses. This process can be carried out several times with minimal losses.

11.2.3 Crystallization at low temperature

For purification of a low melting solid or a thermally unstable liquid, low temperature crystallization is occasionally useful, but this technique is attended by two important complications. First, cooling the materials and apparatus below ambient temperature will cause condensation inside the flask. This may not be a major problem if the compound being crystallized is inert to water, and can be easily dried. The crystallization flask can be fitted with a drying tube, but it is more satisfactory to carry out the crystallization under an inert atmosphere. The problem with low temperature crystallization is that of keeping the solutions cold during the filtration and washing steps. This can be achieved by using apparatus which can be immersed in ordinary cold baths, by using specially built apparatus with integral jackets for holding cooling solutions, or in small scale work by carrying out these steps very rapidly. A number of ingenious solutions to these technical problems have been developed, but only a few methods which involve standard organic laboratory glassware will be described here.

For medium to large scale crystallizations ($\geq$ 1g) a set-up along the lines shown in Fig. 11.3 may be used. The compound is dissolved in the minimum volume of solvent at room temperature and is filtered into a two- or three-necked pear shaped flask. The flask is then fitted with an inert gas inlet and a thermometer adapter containing a filter stick connected to a bubbler. With the filter stick held above the solution the flask is purged with inert gas and placed in a cooling bath. It should be cooled slowly by gradual addition of the cooling agent to the solvent. When crystallization is complete the bubbler is disconnected and the filter stick is connected to an appropriately sized receiver using chemically inert tubing (Teflon). The filter stick is then lowered into the solution and the mother liquor is forced through into the receiver using inert gas pressure. The thermometer adapter will allow sufficient freedom of movement of the filter stick to pack the crystals to the bottom of the flask and drain off the mother liquor thoroughly. The crystals can then be washed by releasing the gas pressure and adding small amounts of *precooled* solvent *via* the three-way tap, using a cannula. The washings can then be removed using the filter stick as before. The cooling bath can be removed and the crystals isolated and dried in the usual way or the low temperature recrystallization can be repeated in the same flask.

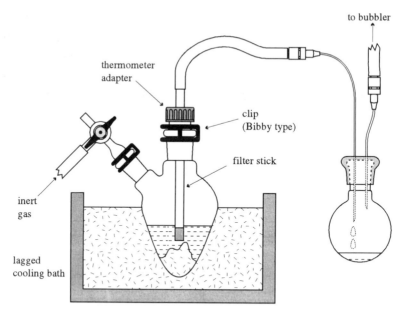

Figure 11.3

The sintered disc of the filter stick should be of porosity 3 or larger in order to avoid blockage. A convenient alternative to a filter stick is to use an ordinary glass rod with filter paper wrapped round the end (Fig. 11.4b). Wrap some Teflon tape round the end of the tube, then carefully fold filter paper over the end and secure it with wire. The Teflon tape will give a better seal because the wire will sink into it. Another alternative, for smaller scale work, is to use a long syringe needle (Fig. 11.4c). Again, the end of the

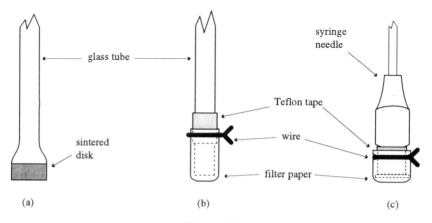

Figure 11.4

needle should be wrapped with Teflon tape and covered with filter paper, then secured with wire. The sharp end of the needle can be inserted into the receiving flask, thus obviating the need for connecting tubing (Fig. 11.5). These devices are also very useful for filtering reaction mixtures under an inert atmosphere, as is the filter flask described in Chapter 6 (Fig. 6.19).

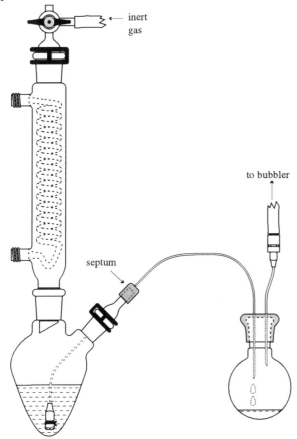

Figure 11.5

Small scale low temperature crystallizations can be carried out in the same device recommended for ordinary small scale work. First of all dissolve the material in an appropriate solvent in the flask. Then purge the apparatus with inert gas and seal the outlet with a small septum (Fig. 11.6). Immerse the flask in a cold bath, and when crystallization is complete, remove the septum and filter the suspension rapidly under inert gas pressure. On a small scale the solution will not warm up significantly during the short time required to filter off the crystals. The crystals can be washed with precooled solvent.

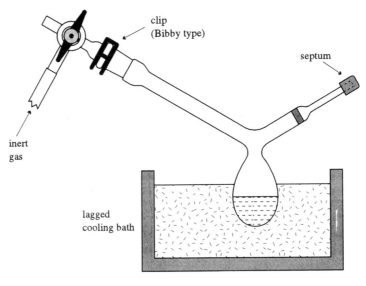

Figure 11.6

11.2.4 Crystallization of air-sensitive compounds

Clearly the crystallization of air and moisture sensitive compounds must be carried out under an inert atmosphere. The trickiest step in this regard is often the introduction of the crude solid into the crystallization flask. This is best avoided if possible by transferring the material in solution. Alternatively, it may be possible to transfer it very quickly in the air or using a "blanket" of inert gas under a funnel (Fig. 6.20). For very sensitive compounds the transfer of the solid must be carried out in a glove bag or a glove box.

Once the solid has been placed in the flask, the methods described above for low temperature crystallizations under inert atmosphere can be used for most crystallization encountered by an organic chemist. The apparatus shown in Figs. 11.3, 11.5 and 11.6 can be used, without the cooling bath, unless it is required.

One particular problem which arises with compounds that hydrolyse on contact with moisture is that small amounts of decomposition products can block filters. Thus, if a sintered disc is used it must be scrupulously dry in order to prevent hydrolysis occurring in the pores of the filter disc, which will rapidly block it. In fact, filter sticks made using filter paper are often better than sintered discs, and at least one dry filter should be kept at hand in case the first one becomes blocked.

11.3 Distillation

Distillation is the most useful method for purifying liquids, and is used routinely for purifying solvents and reagents. Using appropriate apparatus, and some care it can be possible to separate liquids whose boiling points are less than 5°C apart. We will assume that the reader is familiar with the fundamentals of the theory and practice of distillation but it is appropriate to begin by reiterating some basic safety rules.

1. Never heat a closed system.
2. Remember that most organic liquids are extremely flammable so great care must be taken to ensure that the vapour does not come into contact with flames, sources of sparks (electrical motors), or very hot surfaces (hot plates).
3. Never allow a distillation pot to boil dry. The residues may ignite or explode with great violence.
4. Beware of the possibility that ethers and hydrocarbons may be contaminated with peroxides (Section 4.4). Be particularly careful when distilling compounds prepared by peroxide and peracid oxidations and always take precautions to remove peroxide residues prior to distillation!
5. Carry out a safety audit on the compound you plan to distil to check that it is not thermally unstable. Some types of compounds, e.g. azides, should *never* be distilled.

11.3.1 Simple distillation

The conventional apparatus for simple distillation is shown in Fig. 11.7. It is only useful for distilling compounds from involatile residues, or for separating liquids whose boiling points differ by at least 50°C. Moisture can be excluded by attaching a drying tube to the vent or by connecting the vent to an inert gas line. It is essential to add some boiling chips, or stir the liquid, in order to prevent bumping, especially if a finely divided solid, such as a drying agent, is present.

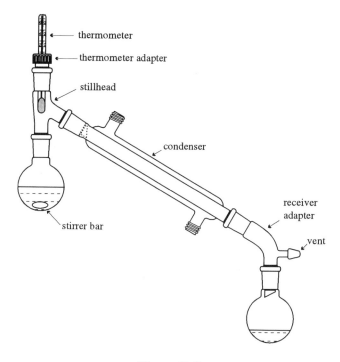

Figure 11.7

The flask should be heated in a water bath, or an oil bath, ***not*** with a Bunsen burner or a heating mantle. The temperature of the bath should be increased slowly until distillation begins and then it should be adjusted to give a steady rate of distillation. If the boiling point is high (>150°C) the stillhead may need to be lagged with glass wool, and an air condenser should be used rather than a water condenser.

For most purposes it is more efficient and convenient to use a compact (short path) distillation apparatus such as that shown in Fig. 11.8. These can be bought, or constructed in two or three convenient sizes to fit the common ground glass joints (see chapter 6 for design details). The main advantage of such devices is that very little material is lost on the sides of the apparatus and joints. Therefore they are particularly appropriate, if not essential, for small scale work.

11.3.2 Distillation under inert atmosphere

If the distillation is being used to dry a reagent, the process must be carried out under an inert atmosphere. The following procedure can be used:

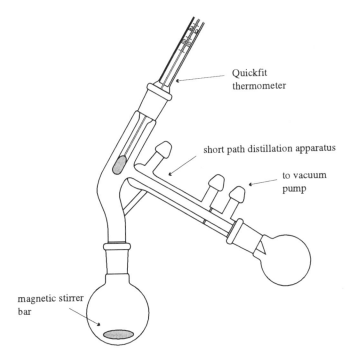

Figure 11.8

1. Dry all the glass apparatus in an oven, or with a heat gun under vacuum, and purge with inert gas whilst cooling. This is most easily accomplished by connecting the apparatus to a double manifold/bubbler system (see Chapter 4). Although Quickfit distillation assemblies can be used, we prefer one piece type apparatus, as shown in Figs. 11.8 and 11.9b.

2. When the glassware has cooled, increase the inert gas flow, quickly disconnect the distillation flask, add any drying agent required, a few anti-bumping granules and the liquid to be distilled, then reassemble the system.

3. Heat the distillation flask *in an oil bath* (do not carry out distillations using a heating mantle) and collect the distillate which comes over at the required temperature.

4. When the distillation is complete remove the collector and seal it quickly with a septum. Most reagents can simply be poured into a reagent bottle before sealing, provided you are quick. However, if the reagent is particularly sensitive to air or moisture a cannulation

technique should be used to transfer it (see Chapter 6). Whatever type of container is used for storage it is always preferable that it be full or nearly so.

11.3.3 Fractional distillation

Separation of liquids whose boiling points are between 5°C and 50°C apart requires the use of an apparatus which gives better contact between the vapour and liquid phases in the distillation column. One of the most common types of fractionating column, a Vigreux column, is shown in Fig. 11.9a. A 50cm long vacuum jacketed Vigreux should allow reasonable separation of compounds which boil 30-40°C apart.

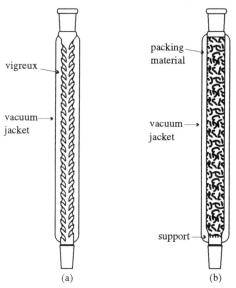

Figure 11.9

One piece assemblies incorporating short Vigreux columns (Fig. 11.10) are somewhat less efficient but very convenient for routine distillations and give good results at reduced pressure.

Efficient separation of compounds with a boiling point difference of 10-30°C can be achieved using a long glass tube packed with glass rings or helices, or for high efficiency, wire mesh rings (Fig. 11.9b). The key to getting good results from a fractional distillation is to raise the temperature very gradually, and collect the distillate very slowly.

Two more sophisticated commercially available designs are the 'Spaltrohr' columns which consist of concentric grooved tubes, and spinning band columns in which a rapidly spinning spiral band is fitted inside the column. Both of these systems have low hold ups and give very high efficiency so they can be used for the separation of small volumes (as little as 1ml) and for separating compounds with very similar boiling points (as little as 3°C). Consult the manufacturer's manuals for operating instructions for these devices.

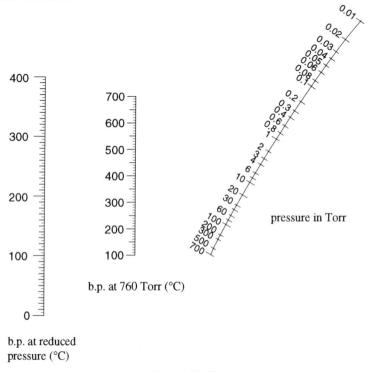

Figure 11.10

11.3.4 Distillation under reduced pressure

Many compounds decompose when heated to their boiling points so they cannot be distilled at atmospheric pressure. In this situation it may be possible to avoid thermal decomposition by carrying out the distillation at reduced pressure. The reduction in the boiling point will depend on the reduction in pressure and it can be estimated from a pressure-temperature nomograph (Fig. 11.11).

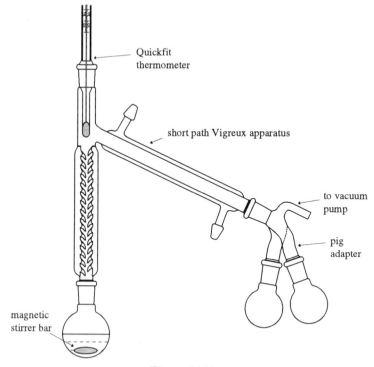

Figure 11.11

To find the *approximate* boiling point at any pressure simply place a ruler on the central line at the atmospheric boiling point of the compound, pivot it to line up with the appropriate pressure marking on the right-hand line, and read off the predicted boiling point from the left-hand line. You can also use the nomograph to find the b.p. at any pressure if you know the b.p. at some other pressure, by first using the known data to arrive at an estimate of the atmospheric boiling point. Note that although pressure is usually measured in millimetres of mercury (mmHg) it is often quoted in different units, especially Torr. Happily 1 Torr = 1mmHg. Boiling points measured at reduced pressure may be expressed in several ways, e.g. 57°C/25mmHg or b.p.$_{25}$ 57°C. As a very rough guide a water pump (ca. 15mmHg) will give a 125°C reduction in boiling point and an oil pump (ca. 0.1mmHg) will give a reduction of 200-250°C.

A typical vacuum distillation apparatus is shown in Fig. 11.10. The chief difference from the simple distillation apparatus is in the design of the receiver adapter. This must allow several fractions to be collected without needing to break the vacuum. The simplest design is the 'pig' type shown in

Fig. 11.12a. When using a pig, remember to grease the joint lightly, around the outer edge only, so that the receiver can be rotated while under vacuum. Also fix the receivers securely using clips.

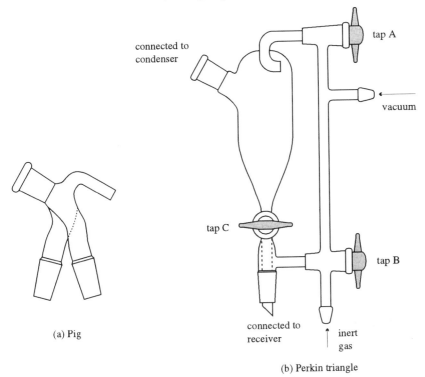

(a) Pig

(b) Perkin triangle

Figure 11.12

A procedure for carrying out a distillation under reduced pressure is as follows:

1. Place the sample in the distillation flask (no more than two thirds full) and add a stirring bar. *Note:* Anti-bumping granules are not effective at reduced pressure and so an alternative must be used. A very narrow capillary which allows a slow stream of air or nitrogen bubbles to pass through the solution is effective, but brisk stirring using a magnetic follower is much more convenient.

2. Assemble the (oven-dried) apparatus, putting a little high vacuum grease on the outer edge of each joint. Ensure that the receiver adapter and the collection flasks are secured using clips, and connect the assembly to a vacuum pump. One convenient method of doing this is to connect it to

a vacuum/inert gas double manifold (Chapter 4). The pump must be protected with a cold finger trap and the line should incorporate a vacuum gauge for monitoring the pressure (Chapter 8).

3. Stir the liquid rapidly and *carefully* open the apparatus to the vacuum. Some bumping and frothing may occur as air and volatile components are evacuated. If necessary adjust the pressure to the required value by allowing inert gas into the system *via* a needle valve.

4. Heat the flask slowly to drive off any volatile impurities and then to distil the product. Monitor the stillhead temperature and collect a forerun and a main fraction, which should distil at a fairly constant temperature. If fractionation is required you may have to collect several fractions and it is very important to distil the mixture slowly and steadily.

5. Stop the distillation when the level of liquid in the pot is running low, by removing the heating bath.

6. Isolate the apparatus from the vacuum and carefully fill with inert gas. If you are using a double manifold, this can be done by simply turning the vacuum/inert gas tap. The flask containing the distillate will be under a dry, inert atmosphere and should be quickly removed and fitted with a tightly fitting septum .

7. Switch off the pump and clean the cold trap.

Using a Perkin triangle to collect fractions

A Perkin triangle (Fig. 11.12b) is a more convenient device for collecting fractions on a larger scale. It is operated as follows:

1. Attach a receiver to the Perkin triangle and open both the apparatus and the receiver to the vacuum using taps A and B.

2. When distillation begins open tap C to collect the forerun.

3. Close tap C so that the next fraction will collect in the bulb, whilst the receiver is being replaced. Use tap B to allow inert gas/air into the receiver, and then replace it with a new one.

4. Close tap A to temporarily isolate the still, then evacuate the receiver by opening tap B to the vacuum. When the pressure has steadied open tap A again. Then tap C can be opened to allow the distillate to drain into the new receiver.

4. Continue to collect fractions by repeating steps 3 and 4 as necessary.

11.3.5 Small scale distillation

The chief difficulty with small scale distillations is that a significant proportion of the sample may be lost in 'wetting' the surface of the column and the condenser. This problem is reduced by using very compact one-piece short-path designs, but this reduces the fractionating efficiency of the columns, leading to less effective separation. Different apparatus designs, giving different trade-offs between recovery and efficiency, are available. A typical example, featuring a short Vigreux and a rotary fraction collector, is shown in Fig. 11.13. Spinning band and Spaltrohr columns combine high efficiency with low hold up so they provide an effective, if expensive, solution to the problem of fractionation of small volumes (> 1ml).

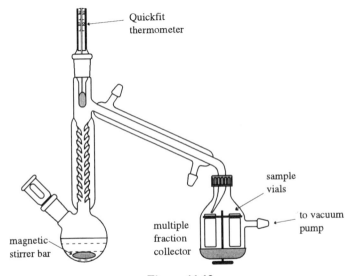

Figure 11.13

Another popular method for distilling small quantities is to use a Buchi Kugelrohr apparatus (Fig. 11.14). The key features of this system are: (i) a horizontal glass oven with an iris closure, which heats the flasks efficiently; (ii) short-path distillation between a series of bulbs which can be moved horizontally in and out of the oven; (iii) a motor to rotate the bulbs. The Kugelrohr can be used to distil very small quantities (< 100mg) and can be operated at high vacuum. A simple distillation is carried out as follows:

1. Place the sample in the end bulb using a Pasteur pipette. If necessary wash the sample into the bulb and then evaporate the solvent on a rotary evaporator.

2. Add two more bulbs and connect them to the straight length of tubing from the motor assembly.
3. Slide the end bulb into the oven and gently close the iris type seal around the connecting joint.
4. Set the bulbs rotating (to prevent bumping and speed up the distillation),
5. Apply the vacuum gradually to prevent frothing and raise the oven temperature gradually until distillation begins (usually indicated by a mist appearing in the first bulb). Adjust the temperature so that distillation continues at a steady rate.
6. When distillation is complete allow the flasks to cool under an inert atmosphere and then remove the bulbs from the apparatus and recover the distillate. If you do not want to wash the collection bulb with solvent, simple clamp it in a vertical position and allow the contents to drain into a flask or sample vial.

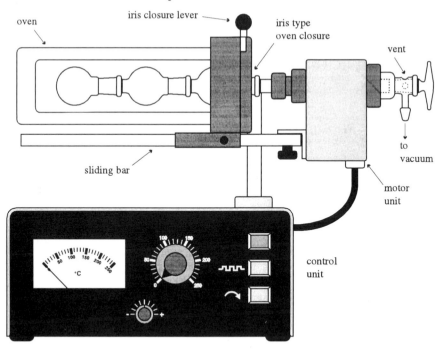

Figure 11.14

Fractionation of compounds with a 20-30°C boiling point difference can be achieved by inserting all the bulbs except one into the oven, distilling the most volatile component into that bulb, withdrawing the next bulb from the oven and distilling the next fraction into that, and so on.

11.4 Sublimation

Sublimation is an excellent method for purifying relatively volatile organic solids on scales ranging from a few milligrams to tens of grams. At reduced pressure many compounds, especially those of low polarity, have a sufficiently high vapour pressure that they can be sublimed, i.e. converted directly from the solid phase into the vapour phase without melting. Condensation of the vapour then gives purified solid product provided, as it is often the case, that the original impurities were much less volatile. Two designs of sublimation apparatus are shown in Fig. 11.15.

The larger sublimator (Fig. 11.15a) consists of a tube with a side arm which is fitted with a cold-finger condenser, and it is used as follows:

1. If the crude material is a solid, powder it and place it in the bottom of the outer vessel. If it is waxy or oily, wash it into the tube with a *small amount* of solvent, cover the side arm with a septum, and remove the solvent on a rotary evaporator.

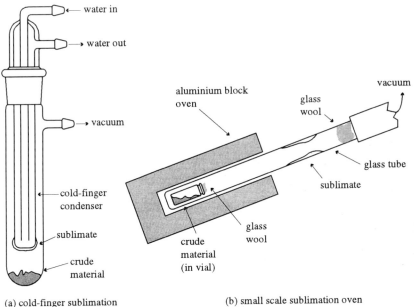

(a) cold-finger sublimation   (b) small scale sublimation oven

Figure 11.15

2. Put some vacuum grease on the joint of the cold-finger condenser and fit it into the sublimator (there should be a gap of approximately 1cm between the solid and the condenser).
3. Evacuate the apparatus slowly to prevent any spattering of the solid. Turn on the condenser water and slowly heat the base of the sublimator.
4. A fine mist of sublimed material on the condenser indicates that sublimation is beginning and the temperature should then be held fairly constant until the process is complete.
5. When sublimation is complete product may be clinging precariously to the cold finger so proceed with great care. Turn off the water, *carefully* allow air/inert gas into the sublimator, and *very carefully* remove the cold-finger. Scrape off the product with a microspatula.

The simple glass tube shown in Fig. 11.15b functions in a similar way except that the water condenser is omitted. Place the crude sample in a small sample vial, put a plug of glass wool in the neck, and drop the vial into the tube. Put a plug of glass wool in the neck of the tube and attach it to a vacuum line. Evacuate carefully and heat the base of the tube gently in an oil bath or in a custom designed metal block as shown. Proceed as before and isolate the sublimed product by cutting the tube and scraping out the solid.

11.5   Chromatographic purification of reaction products

Chromatographic techniques for analysis and purification of reaction products are probably the most universally important of all the skills in which an organic chemist requires expertise!

Before attempting any of the chromatographic separation techniques described in the rest of this chapter you will need to be confident about your ability in running analytical tlc (see Chapter 9), as the two skills are very closely interlinked. For routine separation and purification of reaction products, some form of column chromatography is normally employed. Traditionally long columns were filled with silica and a good head of solvent was used, so that a flow was achieved by force of gravity. This method of chromatography is very slow and the slow elution rate not only wastes time, but also but leads to band dispersion. This *reduces* the resolution and often leads to a large number of fractions which contain a mixture of compounds. The technique has therefore become obsolete, and has been largely superseded by flash chromatography.

11.6 Flash chromatography

Flash chromatography was introduced by Still, Khan and Mitra in 1978.[1] It has cut down the time taken for routine purification of reaction mixtures considerably and, perhaps more importantly, has provided the chemist with a fast and simple technique for separating isomeric materials which have similar polarities. It has become the standard method of purification in many synthetic laboratories, with a resultant reduction in the time taken to achieve many synthetic goals.

The key to the effectiveness of flash chromatography is that a comparatively fine silica powder is used, with a relatively narrow particle range (e.g. Merck C60, 40-63μm, type 9385 or May and Baker Sorbsil C60 40-60μm). This silica gives better surface contact, and therefore more effective adsorption, than that previously used for gravity columns. The pressure required to drive the solvent through increases the resolution by cutting down band dispersion and considerably reduces the time required for running the column.

There are some practical difficulties with the original published method for running flash columns and modifications have therefore been introduced over the years. These make the technique easier to carry out and somewhat safer. This section will lead you through the basic steps for running a modified version of flash chromatography which we find to be very convenient and effective. Flash chromatography is a very powerful and rapid technique for separation of organic compounds, but like all chromatographic techniques *you will only gain expertise by experience,* and these instructions can only be considered as guide-lines. Every separation is different so do not be despondent if you have some failures at first. With practice you should be able to acquire the skill and intuition required to get good separation, even in the most difficult circumstances, every time, *and quickly!*

11.6.1 Equipment required for flash chromatography

The set-up originally described by Still requires the use of long columns, to

1. W.C. Still, M. Khan and, A. Mitra, *J. Org. Chem.,* 1978 **43**, 2923

accommodate a reasonable supply of solvent, but these are difficult to load
and need to be dismantled for addition of extra solvent during the run. We
therefore recommend the use of shorter columns plus a reservoir. It is
important to have a familiar set of columns on hand and we suggest that
you have a set of about five columns, ranging in diameter from about 5mm
to 50mm. A very convenient length for all columns is 25cm, except for
those with a very narrow bore, which can be shorter (Fig. 11.16). It is
convenient to standardize on B24 (24/40) joints for all columns and
reservoirs, so that they are interchangable. A glassblower can make
reservoirs from a round-bottom flask and a male joint. We use a 500ml
reservoir more than any other, but it is also useful to have 100ml, 250ml,

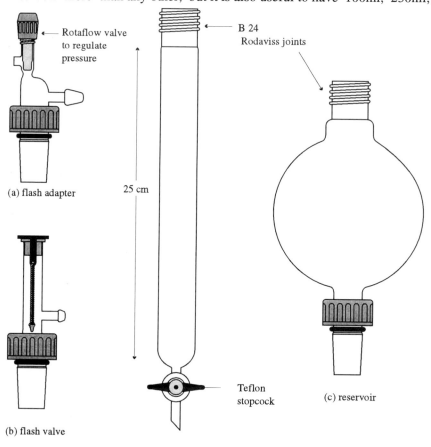

(a) flash adapter

(b) flash valve

25 cm

(c) flash column

B 24
Rodaviss joints

(c) reservoir

Figure 11.16

and 1 litre sizes available. Although equipment fitted with ordinary Quickfit joints, secured by rubber bands, can be used, equipment fitted with Rodaviss joints is recommended. A standard male joint is secured into a Rodaviss socket by a screw collar and rubber O-ring. This is a far superior method for connecting glassware for flash chromatography, in terms of both convenience and safety.

It is difficult to control the column pressure using a gas cylinder and Rotaflow valve, as suggested in the original paper, and this can be dangerous if too much pressure is inadvertently applied. Various alternative methods are used successfully in our labs. We find that the most convenient and safest method is to use a small (fish tank type) low pressure diaphagm pump. If an appropriate model is used it will give a good flow rate without causing too much pressure to build up. We use a Charles Austen Model Dymax 2 pump, but several others are available through laboratory equipment suppliers and they are cheap enough for each worker to have his/her own. Another safe and simple method for pressurizing columns is to use a rubber bellows. The only problem with this technique is that it is difficult to keep the pressure constant, particularly with large columns.

If compressed gas is used to pressurize flash columns we recommend that a some sort of pressure-release valve be incorporated, to avoid unsafe high pressures building up. A simple pressure-release valve, such as the 'flash valve' shown in Fig. 11.17, can easily be constructed and is very effective. The glass part of the 'flash adapter' is made from a B24 joint and

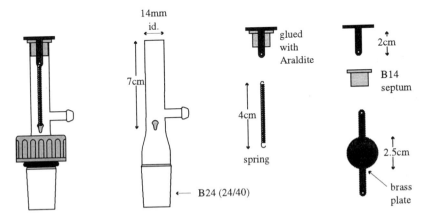

Figure 11.17 - construction of a 'flash valve'

a piece of 14mm i.d. heavy wall tubing. The valve is simply a septum, with the skirt cut off, glued to a shaped metal (brass) plate. The adapter should be constructed with the dimensions shown and fitted with standard 4cm stainless steel lab springs and it will then provide a constant, safe pressure, which will give a suitable flow rate for all purposes.

For analysis of the column fractions, have a tlc tank close at hand containing an appropriate solvent, and several tlc plates, cut to about 4cm wide.

For relatively small columns the most suitable way for collecting fractions is in a rack of tubes and the best type of racks are those which hold the tubes very close together (for example those supplied by Gilson/Anachem for their fraction collectors). Using this type of rack, the column flow can be changed from one tube to the next simply by moving the whole rack. For large columns, conical flasks are often more convenient for collecting fractions.

11.6.2 Procedure for running a flash column

Safety note:

Silica dust is very toxic if inhaled, you should therefore take precautions to avoid breathing it in and always handle it in a fume cupboard. Large volumes of solvent are also used for chromatography and you should take precautions to avoid breathing in the vapours or exposing them to sparks.

1.   Decide on the solvent system and on the quantity of silica

i)   Run tlcs to find a solvent system which will give a good separation of the components of the mixture, and a R_f value of ca. 0.2-0.3 (usually start with ethyl acetate - petroleum ether mixtures). If spots are running close together an R_f value of 0.2-0.3 at the mid-point is normally satisfactory, but if they are well separated, a solvent which puts the lower spot at R_f 0.2-0.3 will usually work. If you know which spot you are most interested in, try to bias your judgement towards this. There are often irrelevant impurities which are either very polar or very non-polar and these can be largely ignored.

ii)   You should try to use as little silica as possible since it is quite expensive. Where the component you require is well separated from other components, a ratio of ca. 20:1 (silica:mixture) should be sufficient. For more difficult separations up to 100:1 ratio can be used (with a ratio of 100:1 spots that are touching on tlc should be separable in the appropriate solvent system). As a rough guide some representations of tlcs are given in Fig. 11.18 which indicate a solvent system which could be selected for separation of the mixture. Below the drawings, an approximation of the relative quantity of silica (compared to the sample) required to bring about separation is given. An important fact to remember for all chromatographic separations is that the more spots there are in the mixture, the greater the ratio of silica for each individual separation. Thus, you normally require *less* not more silica to separate a 3 spot mixture, than for a 2 spot mixture having similar separations. It is in this first step of the procedure that

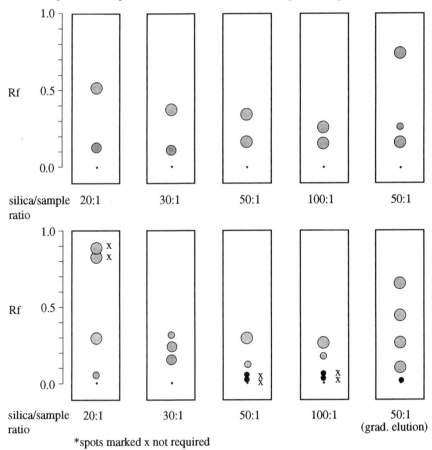

Figure 11.18

the greatest skill and judgement is required and careful attention should be paid to trial tlc's. With experience, the appropriate column and solvent to choose will become almost instinctive.

2. *Choose and prepare the column*

i)   Weigh out the required quantity of silica in a conical flask and make it into a mobile slurry with some of the chosen solvent.

ii) Choose a column which will fill to about 18cm with the amount of silica being used (it is useful to mark columns once you know how much they hold).

iii) To plug the bottom of the column: roll a piece of cotton wool between the fingers so that it is just wider than the column outlet; connect the column to a low vacuum line with the tap closed; drop the cotton wool ball to the bottom of the column, then open the tap. This is the most reliable way to insert the plug.

iv) Mount the column vertically using a clamp stand, pour about 8cm of the solvent in and then carefully sprinkle a layer of fine sand (ca. 1mm), to cover the plug.

v)   *Very carefully* add the silica slurry to the column, in small portions, *via* a powder funnel. Between each portion, pressurize the column to pack down the silica and remove excess solvent. *Be careful not to allow the the solvent to drop below the level of the top of the silica.*

vi) When all the silica has been loaded, leave a good head of solvent on top of it and sprinkle in enough sand to cover the surface of the silica evenly (ca. 1mm). Then force the excess solvent through the column until there is just a small layer left above the sand. There should now be a *flat* layer of silica covered by a thin and even layer of sand.

Note: The column once prepared can be left for a *very short* time, but once you load the sample the remaining steps should be carried out as swiftly as possible, and the solvent flow should be continued without interruption if possible, especially during the early stages of elution. Leaving the column standing with the sample loaded leads to band dispersion and loss of resolution. So, before starting the remainder of the procedure make sure you have everything that you will need to hand, including plenty of solvent.

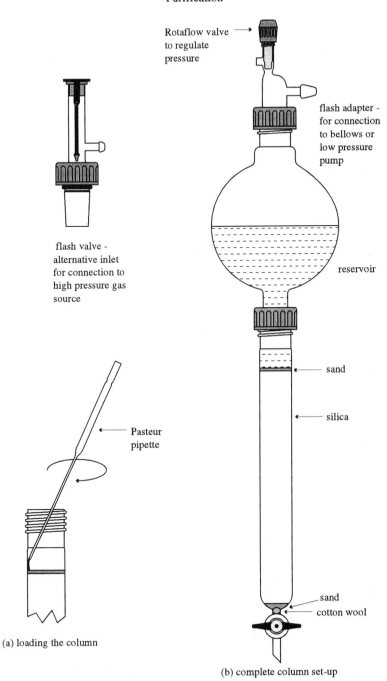

Rotaflow valve
to regulate
pressure

flash adapter -
for connection
to bellows or
low pressure
pump

reservoir

flash valve -
alternative inlet
for connection to
high pressure gas
source

sand

silica

Pasteur
pipette

sand
cotton wool

(a) loading the column

(b) complete column set-up

Figure 11.19

3. *Load the sample*

i)   Dissolve the sample in the minimum amount of solvent, preferably the same solvent that you intend to run the column in (if this is not possible as is often the case, dissolve the sample in a small amount of dichloromethane).[†] Keep a tlc sample of the sample mixture to compare with the column fractions.

ii) Load the solution onto the top of the column, *very carefully*, using a Pasteur pipette to drip it around the walls of the column, just above the sand. Caution must be taken not to disturb the layer of sand. Repeat the procedure using the minimum quantity of solvent to rinse any remaining sample from the flask.

iii) Once all the solution has been added allow the level of liquid to drop so that the top of the sand is just starting to dry.

4. *Add the solvent:*

i)   Add the solvent to the top of the column, *very carefully* at first, again using a Pasteur pipette to drip solvent around the walls of the column just above the sand, and taking care not to disturb the sand. Once a head of solvent is present, more can be poured in *carefully* from a beaker.

ii) Attach a solvent reservoir to the top of the column and secure it using a Rodaviss collar or elastic bands (Bibby clips are not strong enough), and fill with solvent.

5. *Run the column*

i)   Connect the flash adapter to the top of the reservoir and secure it (your set-up should then look like the diagram shown in Fig. 11.19).

ii) Apply pressure to give a fast solvent flow rate and collect fractions continuously. It is very important to maintain a fast flow rate through the column - the solvent should run, rather than drip! *A slow flow rate causes reduced resolution, NOT improved separation.*[††]

[†] For some very non-polar compounds, even small quantities of dichloromethane may cause elution problems and an alternative method of loading can be used. Thus, a solution of the sample is added to a small amount of silica in a round bottom flask and the mixture is evaporated to dryness. The dry impregnated silica is then added to the top of the pre-packed column.

[††] A pressure of 7psi is recommended in Still's original paper, but we tend to rely on flow rate, rather than pressure.

The size of fractions will depend mainly upon the size of the column and as a rough guide, they should be in ml about half the weight of silica (i.e. for a 30g column you should collect ca.15ml fractions). It is a myth that you get less mixtures by collecting smaller fractions - the mixture simply appears in more tubes and leads to a good deal of extra work! You may feel safer collecting relatively small fractions at first, but as you become more experienced you will tend to collect larger fractions, and thus considerably lower the amount of time you spend on chromatography. *Always be careful not to let the column run dry.*

iii) Monitor the column fractions by running tlcs whilst the column is running (5 or 6 spots per tlc plate is usually a convenient amount). You should have time to apply a spot of the previous fraction to a tlc plate whilst the present fraction is collecting.

6. *Analyse tlcs and combine fractions*

   When all the compounds you are interested in have eluted, and you have identified in which fractions they are, combine the fractions as appropriate (keeping fractions that contain mixtures separate from those containing pure materials). Remove the solvent from the combined fractions on a rotary evaporator, and finally remove the last traces of solvent under high vacuum.

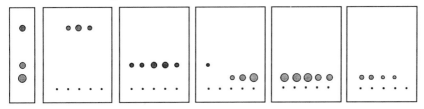

Figure 11.20 Ideal set of tlcs for a successful column

Gradient elution

When the components of a mixture run close together, a single solvent system which gives the spots a tlc R_f of 0.2-0.3, will be effective. However, when the spots are a long way apart, increasing the solvent polarity as the column is running will save a good deal of time and solvent - this is referred to as gradient elution. You should be confident with single solvent chromatography before you attempt gradient elution, as it requires quite a bit of experienced judgement. The procedure for doing this is:

1. Start running the column in a solvent which will give the highest running spot an R_f of 0.2-0.3.

2. When tlc analysis indicates that this component is almost completely off, change the solvent polarity to that which gives the second spot an R_f of 0.3. In some cases you may need to change the solvent polarity in steps.

3. Continue this process until all the spots that you require are off.

11.6.3 Recycling procedure for flash silica

Flash silica is quite expensive but if you can buy it in bulk (25kg at a time) the cost will be reduced by nearly half. Another way to save money is to recycle it, using the procedure given below. The material generated by this method is very reliable.

1. *Washing*

    Suspend 1kg of silica in 1.5 litres of acetone, stir, then filter on a Buchner funnel and wash with an additional 1 litre of acetone, followed by 2 litres of deionized water

2. *Ignition*

    Place the silica in a large crucible and dry at 120-140°C overnight. Then heat in a muffle furnace at 500-600°C for 4h.

3. *Coarse particle removal*

    After cooling, place the silica in a bucket containing about 5 litres of deionized water. Mix the suspension thoroughly, avoiding vortexing, and allow the mixture to settle for 30 s. Then *carefully* pour the top 50% of the mixture into a second bucket. Make the first bucket up to 5 litres, stir again, allow to settle for 30 s and again decant off the first 50% into the second bucket. Repeat this procedure twice further then discard the remaining contents of the first bucket.

4. *Fine particle removal*

    The second bucket should now contain about 10 litres of silica/water mixture. Stir this thoroughly and allow it to settle for 4 min. Decant off the first 50% *carefully* and discard it. Stir again, then allow to settle for 4 min, decant off the first 50% and discard again. Repeat this procedure twice more, after making up to 5 litres each time.

5.   *Activation*

Filter the silica in a Buchner funnel then dry in an oven at 120-130°C overnight. Recovery should be 650-800g.

11.7 Dry-column flash chromatography

This technique was developed by Harwood[2] and can be used as an alternative to flash chromatography. The apparatus simply consists of a parallel-sided vacuum filter funnel, incorporating a porosity 3 sinter (see Chapter 4) and a flask (Fig. 11.21).

11.7.1 Method for running a dry flash column

1.   *Column packing*

Fill the funnel to the lip with tlc grade silica (e.g. Merck Kieselgel 60) and tap gently, then apply suction from a water aspirator, pressing the silica down starting at the circumference and working towards the centre. Continue until a level and firm bed is obtained and there is a head space for sample and solvent addition. The approximate funnel sizes, compared with the quantity of sample to be applied, are given in Table 11.1.

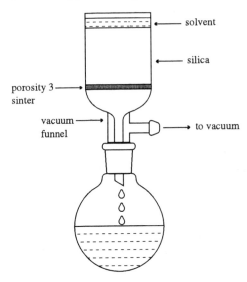

Figure 11.21

2. L.M. Harwood, *Aldrichimica Acta*, 1985, **18**, 25

Table 11.1

Funnel Diameter(mm)	Funnel Length(mm)	Weight of Silica(g)	Weight of Sample(mg)	Fraction Size(ml)
30	45	15	15-500	10-15
40	50	30	500-2000	15-30
70	55	100	1000-5000	20-50

2. *Pre-elution*

Under vacuum, pre-elute the column with a solvent which will give a tlc R_f of about 0.2 for the least polar constituent of the sample. If the silica has been packed correctly, the solvent should run down the column with a horizontal front, but if it channels, the column should be sucked dry and re-packed. Keep the surface of the silica covered with solvent while pre-eluting until solvent starts collecting, then suck it dry.

3. *Loading the sample*

Load the sample as a solution, in the same solvent as used for pre-elution, in an even layer onto the surface of the silica. Alternatively, if the sample is insoluble in the pre-elution solvent, it can be pre-adsorbed onto a small quantity of silica (see Section 11.5.1, loading a flash column)) which is then spread on the surface of the silica in the funnel.

4. *Eluting the column*

Sequentially add solvent fractions to the funnel according to the quantities indicated in Table 11.1, sucking the silica dry in between each fraction, and keeping fractions separate. For each successive fraction increase the solvent polarity by increasing the proportion of the more polar solvent by about 5-10% (e.g. from 50% EtOAc/50% petrol to 55% EtOAc, for a 10% increase). Analyse the fractions by tlc to determine the locations of the components of interest, but as a rough guide, the solvent mixture which would give the compound a tlc R_f value of about 0.5 will probably elute it.

As with any chromatographic technique expertise will only come with experience, but given that, you should be able to separate quite closely running compounds quickly using this technique.

11.8 Preparative tlc

Until quite recently preparative tlc was used extensively for separating small quantities of compounds of similar polarity, but the technique has largely been superseded by the development of flash chromatography and hplc, each of which can give better resolution. Thickly coated 'prep plates' should definitely be considered a thing of the past, but some people still do like to run large (e.g. 20 x 10cm) analytical plates as prep plates, and about 10mg can be loaded onto this size plate.

A high degree of skill is required to draw a tlc dropper across a line at the base of the plate, applying a very thin and even line of the sample solution, without damaging the silica. In between each application the solvent is allowed to dry before a further application is made, and this process is repeated until all the sample has been loaded.

After running the plate (multiple elution is often required for good separation), it is viewed under a uv lamp and the bands are marked with a fine pencil (it is very difficult to use prep tlc for non-uv active compounds). A sharp scalpel is used to cut the bands and scrape them carefully off the plate and the compound is separated from the silica by washing through a cotton-wool plug with a very small quantity of methanol. After evaporation it is necessary to re-dissolve the compound in methylene chloride and filter again to remove traces of silica.

Great skill is required to run preparative tlcs effectively, and even then the sample obtained is often contaminated with significant quantities of grease and tlc plate binding agent. Since there are now much better modern small scale separation techniques available, we do not recommend preparative tlc (see below for details of preparative hplc and Chapter 12 for more on small scale flash chromatography).

11.9 Medium pressure liquid chromatography (mplc)

There are often times when quite difficult separations need to be performed on a fairly large scale. Even if the mixture will separate by flash chromatography, it may be prohibitively expensive, especially if it is a step which needs to be carried out routinely. This is one occasion when mplc is very useful. The resolution of mplc is somewhat better than flash chromatography, but another important feature is that the columns are reused many times, thus avoiding the expense of throwing away large quantities of silica.

11.9.1 Setting up an mplc system

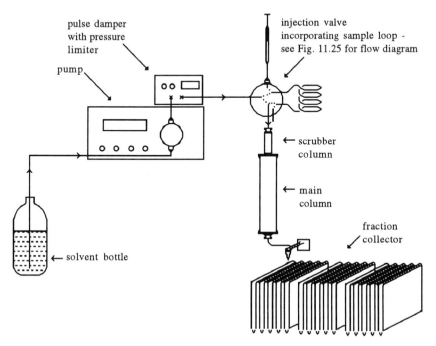

Figure 11.22 (mplc - Schematic diagram)

The mplc system is essentially a simplified and much cheaper version of an hplc set-up. At the heart of the system is any type of pump which will operate at 100psi, with a controllable flow rate of up to 100ml/min. There are a variety of moderately-priced pumps on the market, including one produced by Buchi, which is especially designed for mplc. However, we use a Gilson hplc pump, which has interchangeable flow heads. For mplc, a 100ml head is fitted, but the same pump can also be used for either analytical or preparative hplc, simply by changing the head. Since glass columns are used for mplc it is important that a pressure limiting valve is fitted in-line, to prevent high pressure building up in the column, in the event of a blockage. The valve should normally be set to cut off the pump at about 100psi, but you should consult the instructions for your columns before setting this.

All the pipework used in the mplc system is 3mm Teflon tubing, which is very easily tailored to your needs. The tubing is connected to the various components of the system by plastic screw ferules, which should only be tightened finger tight. It is very easy to make up the tubing to suit your needs, but you will need a flanging tool to fit the ferules onto the tube ends (Fig. 11.23a). The tubing can be connected using threaded plastic sleeves (Fig. 11.23b).

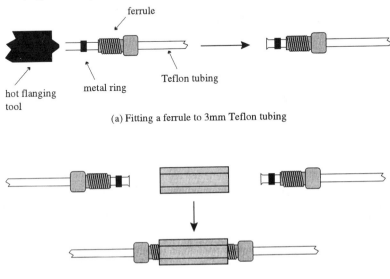

(a) Fitting a ferrule to 3mm Teflon tubing

(b) Connecting 3mm Teflon tubing using a threaded sleeve

Figure 11.23

The sample is introduced into the system *via* an injection valve, but this is a much simpler and less expensive valve than that found in an hplc system. We use a Rheodyne type 50 Teflon rotary valve, which will take a flow rate of 100ml/min (Fig. 11.24). The pump and the column are connected to the red and white connectors respectively, whilst a sample load loop (see below) is fitted to the black and yellow connectors. A Luer fitting is attached to the blue connector, which is where the sample is introduced, using a syringe. The green connector is a vent and should be positioned so that effluent from it can be collected in a beaker or flask.

In the 'load' mode (anti-clockwise position), solvent from the pump inlet (red) passes directly out into the column (white). In this mode the sample inlet (blue) is connected through the load loop (black and yellow) to

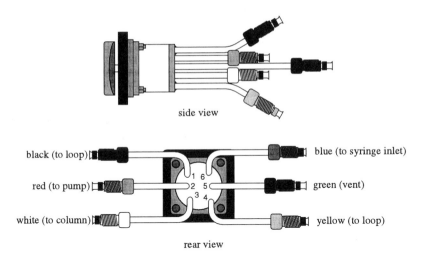

Figure 11.24 Rheodyne Type 50 mplc injection valve

the vent (green). Thus with the solvent flow by-passing the load loop, sample can be injected from a syringe onto the loop, displacing any residual solvent out of the vent. On switching the valve clockwise to the alternative 'inject' position, the solvent stream is diverted through the load loop, introducing the sample onto the column (Fig. 11.25).

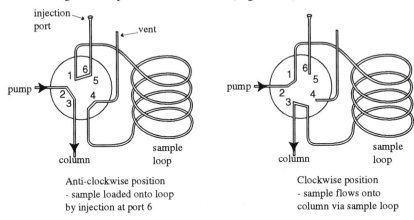

Figure 11.25 Flow through mplc injection valve

Load loops are made very simply by coiling a length of 3mm Teflon tubing which holds the required volume and fitting ferules on either end - 10, 25 and 50 ml loops are useful sizes to make.

There are now several manufacturers of good quality glass mplc columns. The ones we use are made by Buchi and we have found them to be very effective. They are available in a wide variety of sizes and it is convenient to have a selection, but if your budget is tight, just one large column would be very useful. The column we use most frequently is 50cm long by 50mm wide and we can separate between 8 and 20g on this depending on the separation. It is useful to have a small 'scrubber' column in front of the main column, which will prolong its useful life. Each column manufacturer will recommend a packing method, and Buchi supply a special loader for their columns, which is very quick and simple to use, and packs the silica dry under compressed air pressure.

There is a choice of silica available for mplc, but for most purposes ordinary 'flash' (40-60μm) silica is used. It is surprising to find out how much more effective this silica is under mplc conditions than under simple flash conditions. A column packed with 15-20μm silica is more effective for difficult separations but there is a lot to be said for sticking to only one, or perhaps two types of silica and becoming familiar with their characteristics. In most cases getting a good separation is simply a matter of choosing the appropriate size of column and the correct solvent. Other solid phases can also be used on an mplc system alumina, ion exchange resin, Sephadex.

It is very useful to have the mplc system linked to a fraction collector. Any kind of fraction collector will work, but we use a Gilson computer controlled device, which is ideal for a variety of purposes. For small scale work we collect 20 ml fractions (Gilson number 22 rack), and for large scale we collect 60 ml fractions (sintillation rack number 24 with home-made tubes to fit).

A uv or refractive index detector can also be incorporated into the system, or the fractions can simply be analysed by tlc, as for flash chromatography. If the conditions are kept constant the results from mplc are very consistent, so if you wish to repeat separations of the same mixture, it is very easy to predict which fractions will contain each component.

11.9.2 Procedure for using mplc

You will normally have had experience of flash chromatography before attempting mplc, and the two are really quite similar except that the relationship between tlc polarity and column solvent is slightly different.

Again there is no substitute for experience! The following list will guide you through the essential steps.

1. *Decide on column and solvent*

   A solvent which gives an R_f between 0.2-0.4 usually works best, and the size of the column will depend on quantity and separation. As a guide, we routinely resolve 8g quantities of a quite closely running mixture on a 50 x 5cm column, but we have separated up to 20g if the separation is very good. It is always best to rely on experience and for this reason we keep a log of all the runs on the system, which includes tlc sketches, quantities, solvent system, flow rate, column information and comments. Then when you are running something for the first time, you can look back to find similar mixtures which have been separated before and gain some idea of what to use for your sample.

2. *Set up the system*

   Set the system up as described above, re-charging the scrubber column, if one is fitted, and switch on to run solvent through the column. If the column is newly packed, recycle the effluent from the column back into the solvent reservoir and leave the instrument to equilibrate for at least 15 min. If the column already contains solvent, flush this off before recycling. *Also, set the injection valve so that solvent flows through the load loop and flushes away any solvent which was in there!* The optimum flow rate is dependent on column diameter, but will also vary with solvent , separation etc. The following figures are a rough guide: about 20ml/min for a 25mm column; about 45ml/min for a 35mm column; 75-85ml/min for a 50mm column and 100ml/min for anything larger (see under preparative hplc for flow rate calculation). Having chosen the flow rate, set the fraction collector to stop at each tube for an appropriate time.

3. *Sample preparation*

   Mplc is a separation technique and not a clean-up technique, and since the columns are re-used, any base-line material or inorganic solid should be removed from the sample before you start. This can be done either by simple filtration or by filtration through a small pad of silica in a short column. While the mplc column is equilibrating, make up a concentrated solution of the sample in either the chosen solvent or in dichloromethane.

4. *Load the sample*

Switch the injection valve to the 'load' position (Fig. 11.26, solvent by-passing the loop). Using a good syringe with a flat-ended needle, draw up the sample, then invert the syringe and hold it with the needle pointing upwards. Next draw in an air bubble of about 1ml and remove the needle from the syringe. Still holding the syringe in the inverted position, *carefully connect it to the Luer fitting* on the injection valve. Now turn the syringe the right way up and inject the sample until all the solution and a very small bubble have entered the load loop. To avoid a sudden pressure build-up as the sample is injected, reduce the flow rate by half, then switch the injection valve *quickly*. As the sample begins to enter the column, gradually turn the flow rate back up and start the fraction collector.

5. *Running the column*

Whilst the column is running, collect tlcs or use the detector to determine which fractions contain the required components. Unless you are using a refractive index detector, gradient elution is simply achieved by gradually adding more polar solvent to the reservoir.

6. *Finishing the run*

Before you leave the system make sure all the components that were loaded onto the column have been eluted and if necessary, run a more polar solvent through to remove polar material. The column can be left with a length of Teflon tubing attached between the ends to prevent it drying out.

7. *Regenerating columns*

Sometimes a column will become contaminated with stubborn polar material. 'Back-washing' some columns is possible, but this is not necessarily the best way to treat them. It is better to run a polar solvent through, such as methanol (or even water), then gradually change the solvent back to a less polar system, through the sequence: water; methanol; ethyl acetate; ethyl acetate/petrol; petrol.

11.10 Preparative hplc

11.10.1 Equipment required

A preparative hplc system has exactly the same components as an analytical system, except that several features are larger (see Chapter 8). With some systems, such as the Gilson, the pump head can simply be changed to provide the higher flow rates required. The injection valve must also be the large bore type (e.g. Rheodyne 7125) fitted with a large load loop (up to 5ml), and of course the columns are much larger. It is also useful, but not essential, to have a fraction collector, and if this is microprocessor controlled, it can be liked to the detector so that fractions are only collected when a peak is being detected. There are now several extremely sophisticated, computer-controlled systems available, which can be set to inject samples automatically and collect selected peaks into designated flasks. Thus, large quantities of material can be separated over multiple injections. However, for most research purposes the prime requirement is for one-off separations.

The choice of detector can be quite critical. Uv detectors are very sensitive but are of little use if molecules without chromophores are being separated. A refractive index detector is universally applicable but has the drawback that gradient elution is virtually impossible.

11.10.2 Running a preparative hplc column

The sequence of events for running a preparative hplc column is almost the same as that for mplc, as described above. However, there are a few significant differences which should be taken into account, and will be described here.

Care of hplc columns

First of all you should appreciate that preparative *hplc should only be used for separation of clean mixtures.* Some other form of chromatography should always have been carried out previously to remove any material which is significantly more polar than the compounds to be separated and any suspended solid should also be removed. Preparative hplc is a very powerful technique and it is often possible to separate compounds which do not separate on tlc. However, preparative hplc columns are extremely expensive, and although they should last for a very long time, they can easily be ruined by one thoughtless application. Tiny solid particles will damage a column, and for this reason sample solutions should be passed through a fine filter directly before loading

Choice of column, solvent and flow rate

It is very difficult to draw a correlation between the behaviour of a mixture on tlc and the way it runs on hplc, and it is therefore preferable to carry out a few analytical hplc runs to find an appropriate solvent system and determine the quantity which can be separated. If you can use the identical type of column for your analytical work this is of course ideal. There are now sets of columns which run from micro analytical sizes through to large preparative scale, and have uniform characteristics through the series. We use 'Rainin *Dynamax*' columns which have a very sophisticated design and give excellent results for preparative work. The biggest factor which leads to loss of performance in hplc columns is the packing down of the solid phase which occurs with constant use and causes voids, particularly at the ends of the column. One of the features of Dynamax columns is that they are fitted with a compression joint, which closes the voids and enables the columns to perform well over extended periods of use. They can also be fitted with integral scrubber units which extend the life of the main column.

The sizes of the columns increase in regular steps and it is therefore very easy to correlate between the smaller 'analytical' columns and the larger 'preparative' columns. In fact the column which we use for analytical work is 4.5mm x 25cm, and would probably be viewed as a semi-preparative column by an analytical chemist. However, it is very useful for method development, as it takes only a few minutes to equilibrate and to run. Indeed, it can be used as a small scale preparative column for quantities between about 5 and 40mg. For normal analytical hplc the intention is to develop a method which gives the best peak shape with good separation, and column overloading is therefore to be avoided. On the other hand, a good preparative method is one which will separate the components and allow you to 'get away' with maximum overloading, peak shape being largely irrelevant.

For full-scale preparative work either a 22.5mm or a 45mm column is normally used and it is easy to scale up to this size knowing the conditions used for the smaller column. If the columns are the same length the scale-up factor will depend on the area of the column surface, thus:-

$$\text{Scale-up factor, } F \;=\; \frac{\text{Area(large col)}}{\text{Area(small col)}} \;=\; \frac{\pi R^2(\text{large})}{\pi R^2(\text{small})} \;=\; \frac{(D/2)^2(\text{large})}{(D/2)^2(\text{small})}$$

Using these equation, the scale-up factor between the 4.5mm column and the 22.5mm column is 22.35. and between the 22.5mm and the 45mm it is 4. If the columns are not the same length then simply multiply the factor by the proportionate length difference (for example going from a 4.5mm x 12.25cm column to a 22.5 x 25cm column the scale-up factor would be 44.7). To work out the conditions for running the larger column simply multiply the flow rate and the quantity loaded on the small column by the scale-up factor and this should produce identical results on the large column. As a rough guide a 4.5mm x 25cm column runs at about 0.75ml/min and can be loaded with 5-35mg; a 22.5mm x 25cm column runs at about 16ml/min and can be loaded with 100-800mg and a 45mm x 25cm column runs at about 64ml/min and has a capacity of about 450mg to 3.2g. Once you have worked out a good system for prep hplc the run times are quite short, so that large amounts of material can be separated quite conveniently by multiple runs.

Solvents for hplc

All solvents used for hplc must be very high grade and must be degassed before use. Degassing can be achieved by bubbling helium through the solvent *via* a gas diffuser, by standing the bottle of solvent in an ultrasonic bath, or by stirring the solvent under vacuum. The solvent inlet line of the hplc system should always incorporate a filter to prevent small particles entering the system. Ethereal solvent mixtures are not very successful for hplc work and in general it is normally best to use a system where the more polar constituent is also the least volatile.

CHAPTER 12

Small Scale Reactions

12.1 Introduction

For the purposes of this chapter we will define small scale reactions as those involving reaction mixture volumes of less than 5ml. When performing organic reactions on this scale special problems arise, most notably:
1. Difficulties in measuring out small quantities of sensitive reagents.
2. Significant losses of material due to apparatus design.
3. Difficulties in excluding trace amounts of water from moisture-sensitive reactions.

Whenever reactions are performed, material losses are obtained as a consequence of the above problems. Normally these losses only account for a few percent of the total material, however this percentage can increase dramatically as the reaction scale decreases. For example if a moisture-sensitive reaction was carried out on a one mole scale, it would take 18g of water to completely stop the reaction occurring, however if the same reaction is carried out on a 0.1mmol scale, then only 1.8mg of water would completely quench the reaction.

The problem of weighing out small quantities of sensitive reagents is best solved by accurately weighing larger quantities, and making up solutions in an inert solvent, ideally the reaction solvent. Since the molarity of the solution is known, a quantitative aliquot of this solution can then be added to the reaction mixture using a syringe . This effectively reduces the problems of weighing out the material to those that exist in larger scale reactions, and discussed in earlier chapters. As a general rule many of the techniques used in setting up reactions discussed in the earlier chapters, can, with care, be applied to small scale reactions. Indeed there are a variety of miniature chemical apparatus commercially available for just this purpose.

This chapter will outline some of the more specialized techniques that can be employed to alleviate the problems associated with small scale, and that can be carried out without the requirement of relatively expensive specialized glassware.

12.2 Reactions at or below room temperature

When carrying out small-scale reactions at or below room temperature it is quite possible to use a conventional apparatus set-up. Most reactions are carried out in round-bottomed flasks, which are available in sizes down to 1ml. The main problems arise when you come to work up the reaction mixture. If an aqueous or organic extraction is required, then the material must be transferred to a separating funnel. This will inevitably lead to some loss of material during the transfer, but this should be minimal. The main problem arises from the fact that separating funnels rarely come in sizes below 10ml. In addition, because of their design there will always be some loss of material in the apparatus itself. This is mainly because the reaction mixture necessarily comes into contact with a large surface area of glassware, and retrieving material coated over this large area tends to be difficult. A useful alternative for extractions involving 2ml or less is to use a glass sample vial or small test-tube (Fig. 12.1). The reaction mixture can be washed into the sample vial, and the extraction solvent added. The lid of the vial is then put on, and the mixture shaken. Removal of the lid allows any pressure build-up to be released. Particular care should be taken when removing the lid to avoid spillage.

The required solvent layer is then removed using a Pasteur pipette. Three or four extractions are usually sufficient to recover the majority of the material. This technique is particularly straightforward for diethyl ether extractions of aqueous solutions in which the ether layer is required, since it can readily be pipetted from the top of the aqueous layer (Fig. 12.1). If you require the lower solvent layer, this can be recovered either by pipetting away the top layer, or by pipetting from the bottom of the vial. Because efficient separation of the two phases is required, it is preferable to use a tall, thin vial rather than a short, fat one. A second vial containing drying agent can then be used to dry the extracts.

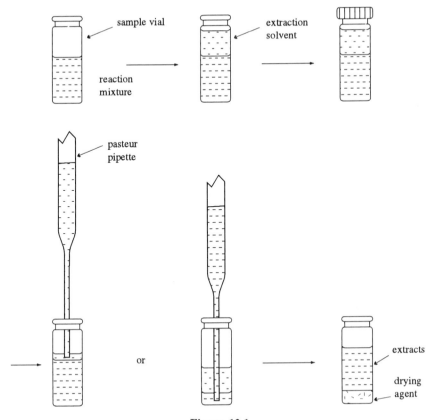

Figure 12.1

Removal of the drying agent is normally achieved by filtration of the solvent mixture, and on a small scale this is best achieved using a Pasteur pipette fitted with a cotton wool plug as the filtration apparatus (Fig. 12.2). Once the solvent has been transferred into the filtration pipette, it can be forced through the plug by applying pressure with a pipette teat. Evaporation of the solvent in the normal way then yields the crude reaction product. As stated earlier, material is inevitably lost with each transfer of apparatus. It is possible to cut down the number of transfers by using the sample vial as the reaction vessel. Very small magnetic fleas are now commonly available, and will fit most small sample vials. Consequently, with magnetic stirring, the vials can serve as small reaction vessels (Fig. 12.3). They are conveniently attached to the top of a magnetic stirrer machine using plasticine.

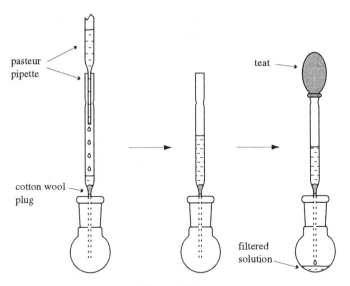

Figure 12.2

When capped, the vial is a sealed system, consequently this set-up is only useful for small scale reactions at room temperature that do not involve an increase or decrease in pressure inside the reaction vessel. For reactions at low temperature, or those requiring a positive pressure of an inert gas atmosphere, it is often more convenient to use a Pyrex test tube fitted with a septum (Fig. 12.3).

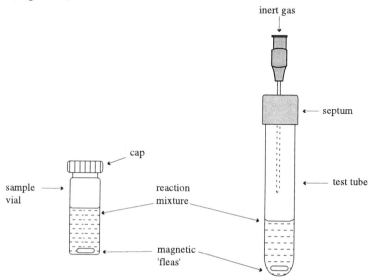

Figure 12.3 Use of sample vial or test tube as a reaction vessel

In all other respects the arrangement is the same and, since Pyrex test tubes are available in a range of sizes, this apparatus can cope with a range of reaction volumes down to about 0.2ml. Again the reaction vessel can also be used for a subsequent extraction procedure simply by replacing the septum with a bung.

12.3 Reactions above room temperature

Carrying out small scale reactions above room temperature, particularly those involving solvent at reflux, is very difficult. The problems are associated with preventing the evaporation of very small amounts of solvent, and material losses in the apparatus. This is usually caused by losses of material through ground-glass joint attachments of condensers to the reaction flask, and by inefficient condensation of the solvent vapour. One solution is to use a sealed tube (see Chapter 9) as the reaction vessel, and this is probably the equipment of choice for reactions involving volatile solvents ($\leq 50°C$) on scales below 1ml. For higher boiling solvents it is possible to use a one-piece apparatus, incorporating an air condenser system (Fig. 12.4). This apparatus can be conveniently constructed at the required size from a piece of Pyrex tubing. The air condenser is adequate enough to prevent the evaporation of higher boiling solvents ($\geq 100°C$). Alternatively,

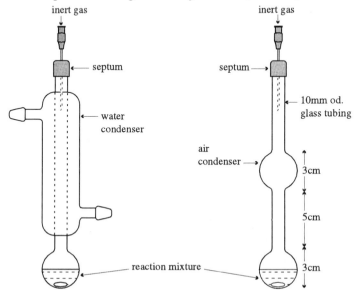

Figure 12.4 Small scale air condenser and water condenser systems.

a one-piece apparatus incorporating a water condenser can also be used (Fig. 12.4). In this case the condenser system is more efficient, however construction of the apparatus is correspondingly more complex. In both cases the reflux apparatus can be easily constructed to allow reaction volumes down to about 0.5ml.

12.4 Reactions in nmr tubes

In many cases it is necessary to monitor closely the progress of a small scale reaction by methods other than tlc; one very useful alternative is nuclear magnetic resonance (nmr). With larger scale reactions this is simply done by removing an aliquot of the reaction mixture and recording its nmr spectrum. Obviously this approach cannot be applied to reactions involving relatively small quantities of material. The answer is to carry out the reaction in an nmr tube. Both 5mm and 10mm diameter nmr tubes can conveniently be used as reaction vessels (Fig. 12.5) in the same way as test-tubes were employed in Section 12.2, but without magnetic stirrer bars.

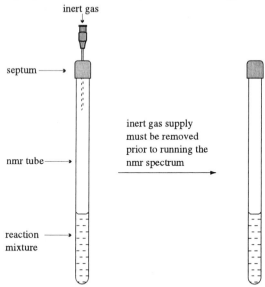

Figure 12.5

The reaction can then be monitored by recording the spectrum of the reaction mixture directly. There are however several important points to note when using nmr tubes as reaction vessels:

1. The reaction solvent should be chosen carefully to ensure that it does not obscure the nmr region to be observed.
2. Magnetic stirring cannot be used because the magnetic flea in the nmr tube would interact with the nmr field causing severe broadening of the spectrum. Similarly the presence of paramagnetic material in the reaction mixture will lead to broadening of the spectrum.
3. For high-quality nmr spectra to be recorded, the reaction mixture should be homogeneous. The presence of solids in the reaction mixture will lead to poorly resolved spectra.

As a general rule you should use the reaction solvent that is normally employed for the type of reaction being carried out, however its deuterated equivalent must be used if ^{1}H nmr spectra are to be observed. Agitation of the reaction mixture is probably best achieved using sonication in an ultrasonic bath (Chapter 9), although periodic shaking will often suffice.

If you require to carry out the reaction at elevated temperatures then it is usually advizable to use a sealed nmr tube. Thick-walled nmr tubes are commercially available if the reaction mixture is required to withstand increased reaction pressures typical of those obtained in sealed tube experiments.

12.5 Purification of materials

The purification of small quantities of materials ($\leq$ 50mg) also poses certain problems. A number of simple techniques used are outlined below.

12.5.1 Distillation

By far the most convenient method of carrying out distillation on a small scale is to use a Kugelrohr apparatus (see Chapter 11). In order to cut down on losses through ground glass joint connections it is often necessary to use a one piece Kugelrohr bulb set (Fig. 12.6). This can be conveniently made to the size required from a piece of Pyrex tubing.

After distillation is complete the apparatus is left to cool, and the purified material recovered by cutting up the apparatus into three sections using a glass knife.

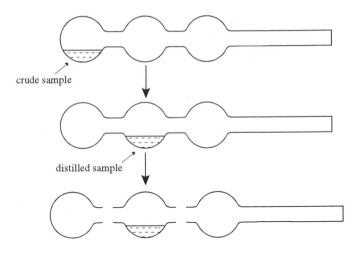

crude sample

distilled sample

Figure 12.6

12.5.2 Crystallization

Crystallizations on a small scale are most conveniently carried out using a Craig tube apparatus as described in Chapter 11.

12.5.3 Chromatography

All the normal chromatography techniques (see Chapter 11) can be used to purify small quantities of material. Indeed preparative hplc and glc are often more successful when small quantities of material are involved.

In the case of flash chromatography however, it is often impractical to simply scale down the equipment. A useful alternative is to employ a Pasteur pipette as the column. Such a column is easily constructed using a pipette containing a cotton wool plug (Fig. 12.7a). The pipette is then filled with the required adsorbent (typically silica gel). The amount of adsorbent used depends upon the quantity of crude sample to be purified, however it is inadvizable to fill the pipette more than three-quarters full, otherwise there is insufficient room for the solvent. Next, the eluting solvent is added to the top of the column and allowed to run through the column under gravity. More solvent is added to the top of the column as required. Once the solvent starts to appear at the bottom of the column, pressure can be applied using a pipette teat, forcing the solvent through at a faster rate. After about two column-volumes of solvent have been passed through the silica it should be ready for use. The sample is applied to the top of the column in the usual way and pressure is applied using a pipette teat. The main difference between this arrangement and the more usual flash chromatography set-up is that the pressure applied to the top of the column is not constant. The teat is constantly being removed to allow the addition of more solvent to the top of the column. Consequently the solvent is not passing through the column at a constant rate. In most instances this does not appear to affect significantly the separations achieved. If, however, this proves to be a problem, a miniature chromatography column with solvent reservoir can easily be constructed from Pyrex glass tubing (Fig. 12.7b). With a cotton wool plug in the bottom this can be used in exactly the same way as the larger columns described in Chapter 11.

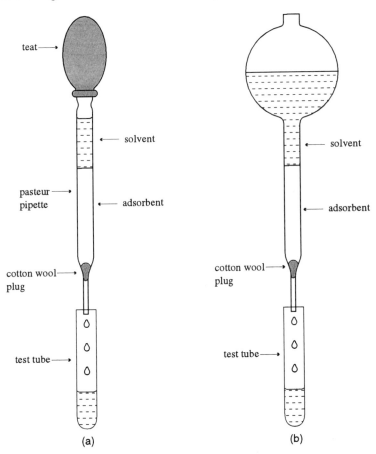

Figure 12.7

CHAPTER 13

Large Scale Reactions

13.1 Introduction

This chapter deals briefly with some of the specialized techniques required when working with larger scale reactions. For the purposes of this chapter we will define a large scale reaction as one involving reaction volumes of between 2 and 5 litres. Working on reaction volumes in excess of this usually requires the use of pilot-plant equipment and is beyond the scope of this book.

When working on larger reaction volumes, several problems arise as a consequence of the scale:

1. The use of syringes to add liquids to a reaction becomes impractical if you require to add volumes of more than 50ml.
2. Stirring the reaction mixtures can become a problem, because magnetic stirrers become ineffective for volumes much above 1 litre.
3. Control of the reaction temperature becomes more difficult as the reaction volume increases, because reaction mixtures will take much longer to heat up or cool down.
4. Exothermic reactions can prove to be a major problem on a large scale since they are prone to induction periods before reaction starts. After the induction period, the reaction can heat up rapidly and, unless extreme care is taken in these situations, the reaction can easily go out of control. It is recommended that very careful monitoring of the temperature is undertaken in such cases, and any addition of reagents which may lead to an exothermic reaction is carried out slowly.
5. Purification of materials on a large scale is often less easily carried out.

On the other hand, some problems that exist with smaller scale reactions can be less troublesome when working on large scale. For example, material losses that occur as a consequence of transferring the material between pieces of equipment become insignificant. Moisture-sensitive reactions are less of a problem because if traces of water get into the reaction they will only affect a very small percentage of the reaction mixture.

13.2 Carrying out the reaction

In most cases large versions of the standard laboratory equipment are adequate. Agitation is almost always achieved using a mechanical stirrer (see Chapter 9), since this is the only device powerful enough to stir large reaction volumes efficiently. As already mentioned, syringes tend to be useless for transferring large volumes of liquid and pressure-equalized dropping funnels serve well as their replacement. A typical set-up for a large scale reflux is outlined in Fig. 13.2. The pressure-equalized dropping funnel can be filled by removing the stopper and pouring in the required material or, if the material is moisture sensitive, then it can be transferred into the dropping funnel *via* a cannula by replacing the stopper with a septum. Heat can be supplied *via* a heating bath or mantle (see Chapter 9) as with other reaction systems, although it is recommended that the heating source be mounted on a lab-jack so that it can be rapidly removed in case of emergency.

If the solution to be added from the dropping funnel requires cooling prior to addition it is possible to use a jacketed dropping funnel, with the cooling mixture (e.g. dry-ice/acetone) placed in the jacket. This is a common set-up encountered for low temperature reactions on larger scales (Fig. 13.1).

13.3 Purification of the products

Many purification techniques are not practical when dealing with large quantities of material. In general the most useful methods of purification that can be applied to large quantities of material are recrystallization for solids, and distillation or steam distillation for liquids. These techniques have already been discussed in previous chapters. Chromatography should be avoided if possible since it becomes a very expensive operation on large scale, but if it is necessary, then medium pressure liquid chromatography (mplc) as described in Chapter 11 is the method of choice.

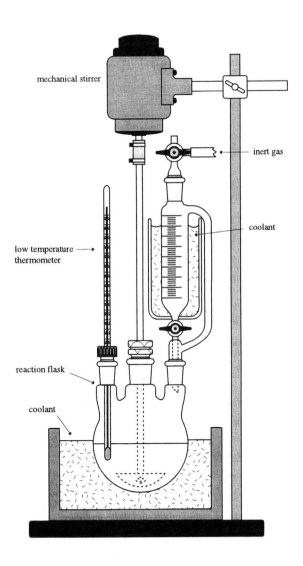

Figure 13.1

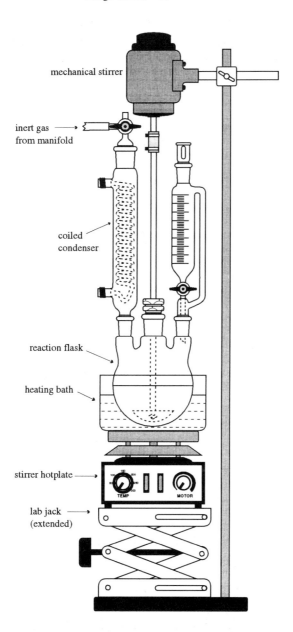

Figure 13.2

CHAPTER 14

Special Procedures

14.1 Introduction

This section deals with some of the specialized procedures which might be encountered. All the topics covered here have one thing in common, namely that the particular type of apparatus used will vary from one laboratory to another and accordingly detailed instructions will not usually be given here for any one piece of apparatus. Representative systems are shown, and common operating procedures discussed. Often there will be someone in the department who is responsible for, or has particular expertise in, one of these techniques. If this is the case always consult this person before you intend attempting the reaction.

14.2 Catalytic hydrogenation

Caution: Extreme care must be taken whenever hydrogen gas and active catalysts are used. Observe local safety precautions strictly.

If you are unfamiliar with catalytic hydrogenation or with the particular local apparatus, do find someone with experience of the apparatus and technique, and familiarize yourself with the manipulations and precautions *before* attempting the reaction.

The reduction of organic compounds using hydrogen and a catalyst is a reaction which is often encountered. Most catalytic hydrogenations are carried out at atmospheric pressure, and organic chemistry laboratories will have their own 'atmospheric' hydrogenation apparatus. These consist of a gas burette (or burettes) connected to a hydrogen supply, and to the reaction flask (see Section 7.4 for details of gas burettes). The detailed operation will depend upon the precise equipment which is used, and a written step-by-step procedure should be available for the set-up. The underlying principle behind most atmospheric pressure hydrogenators is the same and a typical arrangement is shown in Fig. 14.1. Other apparatus might be arranged somewhat differently but the operating procedure for most atmospheric pressure hydrogenators is the same. For the apparatus shown the procedure is as follows:

1. With taps A and B open, raise the levelling bulb to fill completely the gas burette with liquid (usually water, but sometimes mercury), then *close tap B.*

2. Connect your reaction flask using vacuum tubing and *close tap A.* See below for important information on preparing the reaction mixture. It is a good idea to use a long necked reaction flask in case the solvent bumps, and a thin ring of grease should be applied to seal around the top of the connecting joint.

3. Connect the hydrogen supply but keep *tap D closed.* Then connect the vacuum source to tap C and *open tap C* carefully to evacuate the main part of the system. Once the system has stabilized (indicated by the manometer) *close tap C* and carefully *open tap A* partially to evacuate the reaction flask. Then with the hydrogen supply turned on carefully *open tap D* to introduce hydrogen.

4. In order to saturate the system with hydrogen, repeat step 3 twice more.

5. The system should now have *taps A and D open* and *tap C closed*, with the hydrogen supply on. Now slowly *open tap B* to allow the gas burette to fill with gas, adjusting the levelling bulb as necessary. More than the theoretical amount of hydrogen should be introduced if possible.

6. With the gas burette filled, *close tap D*, adjust the levelling bulb so that the manometer mercury levels are equal and record the burette reading.

7. The reaction is then agitated (stirred or shaken) and the uptake of
    hydrogen monitored using the burette. To measure the amount of gas
    that has been absorbed, equalize the manometer mercury levels before
    recording the burette reading.

8. When the required amount of hydrogen has been used or no more is
    being taken-up, purge the system with nitrogen (step 3 using nitrogen
    instead of hydrogen) then remove the reaction flask.

It is important to follow the procedure closely, and to consider the effect of
opening any tap *before* doing so.

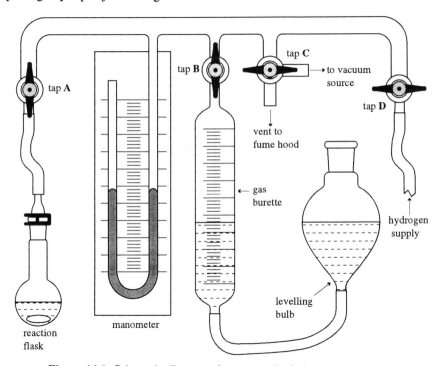

Figure 14.1 Schematic diagram of an atmospheric hydrogenator

When filling the reaction flask, add the catalyst first, followed by the
solvent, and then your substrate. Addition of an active batch of catalyst to
solvent can cause fires. Do make sure that all air is removed from the
system by following the detailed procedure which applies to your particular
system, and when the reaction is over, ensure that as much residual

hydrogen is removed as possible; again, follow the procedure. Ensure that a safety screen is used as much as possible throughout the whole operation.

Filtration of the reaction mixture *must* be done carefully. Filter through a pad of Celite on a sintered glass funnel, but do not allow the catalyst to dry out. Wash through with more solvent and dispose of the wet catalyst/Celite mixture properly. Most laboratories have a special bottle for catalyst residues; use it. Numerous fires have resulted from wet catalyst residues being placed in a waste bin, since the residue dries out in the air and ignites. Raney nickel residues are notorious in this respect. One way to avoid ignition is to centrifuge the reaction mixture and pipette off most of the supernatant. Resuspend this residue by adding more of the reaction solvent, and centrifuge again. This process can be repeated two or three more times as necessary. Do not attempt a catalytic hydrogenation until you know what to do with the catalyst residue.

The above procedure is sometimes inconvenient for small scale work (it depends upon the type of hydrogenation system available), and we often resort to the use of a small balloon of hydrogen connected to the reaction *via* a three-way tap, which is also connected to a manifold. This technique is illustrated in Fig. 14.2.

Fill the balloon with hydrogen (several times to remove air) and connect it to the flask. Connect the three-way tap to the manifold, open to vacuum (carefully) and then to the inert gas (Fig. 14.2a). Several cycles will be required. Then turn the tap so as to isolate the system from the manifold, and allow hydrogen to enter the flask (Fig. 14.2b). The reaction can then be monitored in the usual way. When the reaction is over, vent the excess hydrogen from the balloon (*safely*) to a fume cupboard (Fig. 14.2c). Residual hydrogen can be removed from the system by the use of several cycles of evacuation followed by admission of inert gas using the manifold as above (Fig. 14.2a). If you do not have a manifold available the hydrogen can be introduced from a balloon *via* a syringe needle to a flask fitted with a septum (see Section 9.2.7 for more details about using balloons). *All precautions referred to in the use of the atmospheric hydrogenator must be observed here of course.*

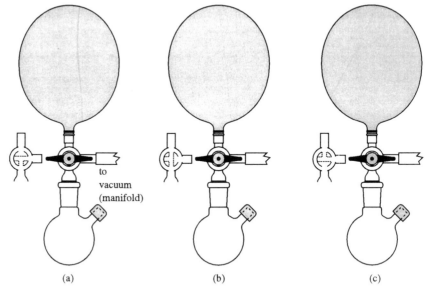

to
vacuum
(manifold)

(a)   (b)   (c)

Figure 14.2   Small scale hydrogenation using a balloon

Medium- and high-pressure hydrogenations require specialized equipment and great care. This equipment usually consists of a metal reaction vessel and the appropriate 'plumbing' to allow its safe pressurization with hydrogen. These reactions are potentially most hazardous, and *must* be carried out under the close supervision of the person who is responsible for the apparatus.

14.3 Photolysis

Caution: Ultraviolet radiation is very damaging to the eyes and skin.

The reactor must be properly screened. Wear special protective goggles, or better still a face shield which offers protection against ultraviolet radiation if the apparatus is to be adjusted (or samples taken) while the lamp is on. When doing this also protect the hands with gloves and make sure that no other areas of skin would be exposed to radiation in

the event of an accident. It is much safer to turn off the lamp when such
manipulations are being carried out.

Preparative photochemical reactions are usually carried out in an
immersion-well reactor, the usual design is shown in Fig. 14.3.

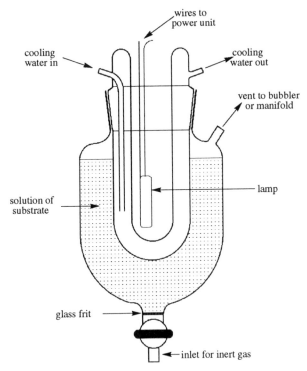

Figure 14.3   Photochemical reactor

Air is removed from the solvent by bubbling nitrogen or argon up
through the solution *via* the sintered glass disk. Ensure that the correct
choice of lamp has been made. Low pressure lamps emit most of their
radiation at 254nm, are low power (up to ~ 20W) and require a quartz
immersion well (not Pyrex). Most preparative reactions use the much
higher power (100-400W) medium pressure lamps, as these emit their
radiation over a much wider range (mainly at ~ 365nm with other bands at
both shorter and longer wavelength). Occasionally a filter will be required
and it is important that the correct one is used (this will be specified in the
preparation being followed).

Reactions are usually run at fairly high dilution (up to ~ 0.05M) and the solvent should be pure and chosen with care. It must not decompose under ultraviolet irradiation and should not absorb at the wavelength being used for the reaction. Work up often involves no more than evaporation of the solvent followed by purification.

14.4 Ozonolysis

Caution: Ozone is toxic, and ozonides potentially explosive.

Ozone is generated using a commercial ozonator (or ozonizer) which can produce a concentration of up to 8% in oxygen, and which will be available in most organic research establishments. The operation of these is very simple providing that the instructions for the particular device are followed carefully. Make sure that these are consulted before attempting the reaction.

The compound to be ozonized is dissolved in the appropriate solvent, and cooled to the desired temperature. Ozone is then passed through the solution until no more starting material remains. For most purposes an excess of ozone can be used. It can be difficult to avoid this, but an indicator which can be added to the solution to show you when there is free ozone in the solution can sometimes be most valuable in avoiding over-oxidation.[1]

The work up depends upon the desired product, but will include a reagent which reacts with the ozonide. This reagent is almost always added in excess, and before any product isolation is attempted. Make sure that you allow plenty of time for the ozonide to react, as isolation of ozonides is to be avoided due to their potential for violent explosive decomposition. Once the ozonide is fully reacted the reaction can be processed in the usual manner.

1. T. Veysoglu, L.A. Mitscher, and J.K. Swayze, *Synthesis*, 1980, 807.

14.5 Flash vacuum pyrolysis (fvp)

This technique is not often encountered in synthetic organic chemistry, but it can prove invaluable in some circumstances. As with most of the topics in this chapter, the exact type of apparatus used will depend on what is available in your department. One simple (schematic) set up is shown in Fig. 14.4. It is important to vaporize the substrate at the appropriate rate, and to make sure that the thermolysis temperature is correct. If this information is not available then some experimentation will inevitably have to be carried out. A typical vaporization rate might be between 0.5 and 1.0g per h.

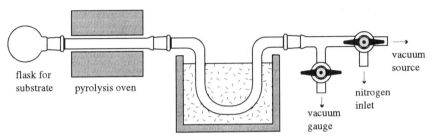

Figure 14.4   Schematic representation of an apparatus for FVP

As with all high vacuum work, care must be taken. After all of the substrate has passed through the hot tube, turn off the furnace and allow to cool to room temperature (still under vacuum). Then turn off the pump and admit nitrogen to atmospheric pressure. Remove the traps to a fume cupboard and allow to warm to room temperature, and work up in the usual way. If the desired product is unstable towards air, water, or is simply very reactive, then a more sophisticated pyrolysis system might be required, and more elaborate work up procedures used.

14.6 Liquid ammonia reactions

Caution: Ammonia is a powerful irritant, toxic, and the gas is flammable. Conduct all reactions in an efficient fume cupboard and avoid all contact with the liquid.

Liquid ammonia (b.p. -33ºC) is a solvent which is not encountered frequently, but which does have several important general uses, in particular 'dissolving metal' reductions ("Birch" type reductions) and most reactions involving lithium amide or sodium amide as bases. Ammonia gas from a cylinder is condensed directly into the flask (Fig. 14.5).

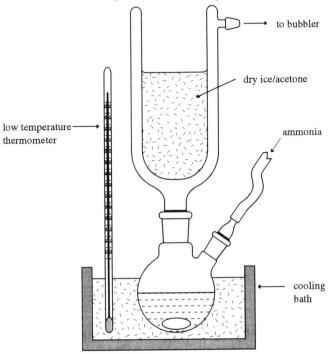

Figure 14.5

The apparatus is set up as in Fig. 14.5 and a rapid flow of ammonia is used to flush out the system. A small volume of acetone (or ethanol) is poured into the condenser, and solid carbon dioxide pellets are added (*very slowly at first*) until the condenser is nearly full. The ammonia will begin to condense, and when the required volume is obtained, the ammonia flow is shut off and the gas inlet replaced by a septum or stopper. If undried impure ammonia will suffice, and often it will, then the reaction is carried out as normal. A cooling bath can be added if a long reaction time is anticipated, or if a temperature below -33ºC is required (see Chapter 9).

If dry liquid ammonia is needed this is usually obtained by distillation off sodium. The appropriate volume of ammonia is condensed as above and small pieces of sodium are added to produce a blue solution. The ammonia can then be distilled using a normal distillation apparatus (Chapter 11) except that the receiver (usually the reaction flask) is cooled in a solid carbon dioxide/acetone cooling bath. The ammonia in the distillation flask must remain blue throughout). The distillation apparatus is disconnected from the receiver which is then fitted with a cold-finger condenser and the reaction carried out as normal. The work up is usually simple. Solid ammonium chloride is added *carefully* and the ammonia allowed to evaporate (Chapter 10). The product may then be isolated and purified in the usual way.

CHAPTER 15

Characterization

15.1 Introduction

This Chapter deals with the type of physical data that are required for the proper characterization of the purified product. No theory is discussed as this is well covered in other sources and, given good data, it is often possible to find a colleague (for example) who will help out if you are unable to interpret a spectrum. With inadequate data it will be difficult to be certain of the structure and purity of your product, and it will certainly be more difficult to interest the colleague referred to above!

It is important to acquire as much information as possible on your product. It might be "obvious" from the nmr that the structure is what you think that it should be, but it is still necessary to record (at least) the ir and mass spectra. These might simply confirm the nmr data, or they might raise other structural possibilities. The collection of physical data has already been touched upon in Chapter 3, and it is advisable to consult this in addition to the current Chapter.

The full set of routine physical data which could and, ideally, should be obtained on a pure compound is as follows; ir, uv, high field nmr (^{1}H and ^{13}C), and low and high resolution mass spectra, m.p. or b.p., microanalysis (for a new compound). If the compound is optically active then the optical rotation must be measured. Only mass spectroscopy and microanalysis from this list are destructive techniques, but modern techniques mean that only a small amount of material need be "sacrificed". Some general points concerning these techniques and the sample requirements are given below.

15.2 Nmr

Nowadays the ^{1}H nmr spectrum is often the first measurement taken. The size of sample required depends on the type of spectrometer used. A continuous wave machine operating at 90 MHz will need at least 10mg of a normal organic compound, probably more. Pulsed Fourier transform spectrometers require less, 5mg being a normal amount, and good spectra can be obtained using much smaller quantities. Most spectra are measured in deuteriochloroform (CDCl$_3$) although other solvents will be required from time to time. A typical solvent volume would be *ca.* 0.4 - 0.5ml in a 5mm tube. Routine measurement of the ^{13}C nmr spectrum usually requires more sample (25-50mg) but good spectra can be obtained on less, it simply takes more time.

Choice of solvent for the nmr spectra is important. Relative chemical shifts are solvent dependent and ideally all spectra should be measured in a standard solvent. Deuteriochloroform (CDCl$_3$) is the generally accepted "standard" solvent and it is advisable to use this where possible. If your compound is not sufficiently soluble in CDCl$_3$ then an alternative must be found. Lack of solubility in CDCl$_3$ is usually due to the polarity of your compound, very polar materials and those with extensive, strong, hydrogen-bonding networks often need solvents other than CDCl$_3$. Such materials usually dissolve sufficiently well in hexadeuteriodimethylsulphoxide (DMSO-d$_6$), tetradeuteriomethanol (MeOH-d$_4$), or deuterium oxide (D$_2$O). All these solvents will absorb moisture readily from the air, and it is advisable to purchase small ampoules and protect them from air once opened. In this way, if the "water" signal should become too great then a new ampoule can be opened and relatively little solvent wasted.

Which solvent to choose depends on several factors. The most obvious of these is solubility, sufficient compound must dissolve to provide a spectrum. Given this then the nature of your compound, and what you need from the spectrum, will be important. Both D_2O and MeOH-d_4 will "exchange out" exchangeable protons such as hydroxyl, amine, or amide protons. If you wish to observe the positions of such protons then it is advisable to try DMSO-d_6. It is also wise to check in the literature for similar types of compounds and see which solvent has been used in the past for running nmr spectra.

If you are seeking to make comparisons of the nmr spectrum of your compound with a known compound, or class of compounds, then it is imperative that nmr spectra recorded in the same solvents must be compared. This also applies to solvent mixtures. Sometimes it is suggested that a small quantity of DMSO-d_6 be added to $CDCl_3$ if your compound is not sufficiently soluble in $CDCl_3$ alone. This is inadvisable if you wish to make comparisons, unless you are careful to note the *exact* ratio of the two solvents. Addition of DMSO-d_6 to $CDCl_3$ results in a change of chemical shift of the resonance for residual $CHCl_3$, a peak which is often used as a reference point in ^1H nmr spectra. The chemical shift of the proton of $CHCl_3$ in $CDCl_3$ is 7.27 δ, in MeOH-d_4 it is 7.88 δ, and 8.35 δ in DMSO-d_6. Be wary of using mixtures of solvents.

If a polar solvent is needed for nmr spectroscopy, then consider also how you will recover the sample if you need to, as DMSO-d_6 is rather difficult to remove without extensive exposure to high vacuum (but prolonged exposure will usually succeed). In this case MeOH-d_4 might be preferable, provided that it is acceptable in other ways (see above).

The solution used must be free of paramagnetic metal ions (it usually is) and particles (it usually is not). Place sufficient sample into a clean, small, vial (weigh it in if you are unsure how much to use), add the solvent (ca. 0.5ml) to dissolve the sample. After taking up this solution into a Pasteur pipette, filtration through a small wad of cotton wool forced into a Pasteur pipette, directly into the nmr tube will usually remove sufficient particles to allow a good spectrum to be obtained. Be very careful that no fragments of glass are broken off the pipette during this filtration, and on removal of the pipette being used as a filter.

The high field ^{1}H nmr spectrum will show up impurities containing protons. Given that the compound has been purified, the most common impurity peaks observed in the ^{1}H nmr spectrum are those from the last solvent used (for example, solvents used in crystallization or chromatography). This should be avoided, and it is always possible unless the boiling point of your product is close to the solvent (which it should not be) or your product is a crystalline solvate (not that common). Thorough exposure to high vacuum should suffice but some very viscous oils and gums will "hold on" to solvent due to the very slow rate of diffusion. If warming in high vacuum fails to remove all solvent, and a "clean" ^{1}H nmr spectrum cannot be obtained, then dissolve the sample in a small amount of CDCl$_3$ and evaporate. Repeat once or twice and most of the residual solvent should be CDCl$_3$ rather than (say) ethyl acetate. This will improve matters, but do not forget that your sample will still be impure on evaporation as it will contain residual CDCl$_3$.

15.3 Ir

The sample again needs to be free of impurities and solvents for infrared spectroscopy. There are various methods for sample preparation and which you choose will depend largely on the type of compound. The amount required is no more than a few milligrams. For a liquid sample the spectrum can be obtained using a thin film obtained by compressing a small drop between sodium chloride plates, or as a solution (usually in chloroform) using solution cells. The spectra of solids can be recorded either as mulls with a hydrocarbon ("Nujol", for example), or by mixing with KBr and compressing to form a thin disk. Which you use will depend upon the facilities available, and often on the usual working practice of your department.

15.4 Uv

Ultraviolet spectroscopy is only of use if your compound has a characteristic chromophore. There is little point trying to measure weak bands which will provide no information. However, it is of considerable value in several areas of research; for example, natural product isolation, heteroaromatic chemistry, porphyrin and related chemistry, and in the study of dyestuffs. The amount of material required is usually very small (fractions of a mg) since the extinction coefficients are usually large. The sample must be as pure as possible and is dissolved in the solvent of choice (usually spectroscopically pure ethanol). The concentration must be known accurately before extinction coefficients can be calculated, and will vary depending upon the type of chromophore. An estimate of the concentration to used can be made if the extinction coefficients of compounds similar to that being studied are available. If this data is not available make up a solution accurately and dilute it (accurately!) until a reasonable spectrum is obtained.

15.5 Mass spectra

There are three pieces of useful information which can be obtained from mass spectroscopy; the molecular mass, composition, and the fragmentation pattern of your compound. The accurate molecular mass is of primary importance since this will confirm the composition of your compound. Fragmentation information might be of value for supporting the proposed structure, possibly by comparison with known compounds. The amount required is minimal (a few mgs at most), and the material should be reasonably pure. If you are unable to obtain good microanalytical data the accurate mass measurement may provide an acceptable alternative.

15.6 M.p. and b.p.

These are usually straightforward. There are various forms of melting point apparatus in widespread use, so check carefully on the procedure appropriate for the apparatus available to you. Always obtain a "rough" melting point before attempting to make accurate measurements and it is often useful to "calibrate" the apparatus by measuring a known (pure!) compound with a similar mp to the product. If there is a significant discrepancy then a more reliable apparatus must be used. Do not forget to get the inaccurate apparatus repaired, and discard it if necessary The compound needs to be pure and free from dust, and the temperature must be raised very slowly as you near the melting point. If there is a range over which the compound melts (there usually is) then record it; do not estimate an "average" reading. If a capillary tube is used, it is sometimes useful to examine the upper part of the tube for sublimate or distilled decomposition products.

If you have distilled your product to isolate and purify it then you should already have the information required for reporting the boiling point. It is important to quote the range of temperature (if observed) over which the compound distils, the pressure (measured as it is distilling), the vapour temperature (if measured), and the bath temperature. All these will be useful when you or anyone else come to repeat the work, and most of this information will be required at some time for a publication, report, or thesis.

15.7 Optical rotation

If your product is, or should be, optically active then the specific rotation will need to be measured and recorded. The precise value of this property is dependent on the wavelength of the light used, solvent, concentration, and temperature. Moreover, great care should be taken to exclude any by-products from the reaction since, although these might be present in small quantities, they might have very large rotations and make your measurements quite misleading. Clearly then, it is important to be sure of the purity of your product, and to make up the solution carefully and accurately. If you are unsure then make a measurement using a known compound before you try to measure the rotation of your product (assuming that the specific rotation of your product is not known). Usually you will use the sodium-D line and measure at ambient temperature, but be sure to record the concentration, solvent, wavelength, and temperature along with the actual value of the measured specific rotation. Occasionally optical rotatory dispersion and/or circular dichroism spectra will be required. These measurements will usually made by specialists and the specific requirements for your particular type of compound are best discussed with them.

15.8 Microanalysis

There are several schools of thought on this topic. Some maintain that all new compounds must be analysed, whereas others say that, with the modern array of physical methods (high resolution nmr spectroscopy and high resolution soft ionization mass spectroscopy, for example) the need for combustion analysis no longer exists. Many follow a middle course and use microanalysis for crystalline compounds which are available in sufficient quantity, and high resolution mass spectrometric measurements in all other cases. The course you take will depend upon circumstances (the requirements of your supervisor or the department, for example).

It goes without saying that the compound must be homogeneous and free from dust, inorganics, etc. For solids, careful recrystallization using pure, filtered solvents followed by equally careful filtration will usually suffice. Ideally, for oils, distillation followed by sealing in an ampoule will provide acceptable samples. However, provided that distilled solvents are used, care is taken in fraction collection to avoid particles of silica in the fractions, and great attention is applied to solvent removal, it is often possible to obtain acceptable microanalysis on samples obtained by flash column chromatography. Solids must be dried in high vacuum in a drying pistol to remove traces of solvent, and submitted in clean, dry vials.

Microanalytical data should be within 0.3 - 0.5% of theory. If this is not the case, then the sample is either impure in some way, or not what you think it is. Try to fit the data to a sensible molecular formula, and if a good fit is found, then try to work out what structure might correspond with this formula and which would be compatible with the rest of the data (nmr, ir, mass spectra etc.). Occasionally compounds will crystallise with a molecule of solvent (or water). Try molecular formulae which include a molecule of the solvent of crystallisation (or possibly water). If this works, then you should be able to detect this molecule of crystallisation in the ^{1}H or ^{13}C nmr. Occasionally, samples will absorb water from the air, and if you and the analyst are unaware of this inaccurate results can be obtained. This hygroscopic behaviour is easily tested by carefully weighing a freshly purified sample and checking if it gains weight after exposure to the atmosphere for a few hours. An ir spectrum "before and after" should reveal any serious contamination with water. This problem can usually be solved by careful sample preparation, isolation, and sealing of the sample vial or ampoule (preferable). Make sure that you provide the analyst with the information regarding the hygroscopic nature of the material.

If you cannot obtain a fit with any reasonable molecular formula, then accept that the sample must have been less pure than you thought. Re-purify another sample and try again.

15.9 Keeping the data

If you work for any length of time in the laboratory you will rapidly acquire a large number of spectra. It is important that you keep these safe and in proper order, with an unambiguous cross-referencing system so that you (and anyone else) can locate the spectra or measurements which apply to the product of a particular experiment (see Chapter 3 for detailed advice on this). Spectra are best kept in clearly labelled folders or binders of some description, preferably ones which allow for removable attachment of the spectra. The other data should be recorded in the laboratory notebook along with the experimental write up. If data sheets are used then all the data should be recorded on these as they are measured.

CHAPTER 16

"Trouble Shooting"; What To Do When Things Don't Work.

Do not despair-yet...

Some reactions will not work, for proper chemical reasons associated with the substrate. These may or may not be "obvious" (many things become "obvious" with hindsight). Do not jump to the conclusion that you have encountered such a reaction. Before you can conclude that this is the case a number of possibilities must be explored.

The first and most obvious is to make sure that the starting material is pure, dry, and free from solvents, and that it is indeed what you think it is. A critical perusal of *all* the analytical and spectroscopic data will usually be sufficient. If you have used several batches of starting material then do make certain that the spectra which you check are from the batch which you used in the failed reaction. If necessary, re-run the spectrum to ensure that the starting material has not partially decomposed, or picked up moisture.

If the starting material was purchased from a chemical supplier, do check that it really is what it purports to be by checking the nmr and ir spectra, and any other data which might be available. If a reagent has been purchased, then it is usually possible to check its reactivity by carrying out a known, reliable, literature reaction. If it is possible to purify the reagent, do so, and make sure that you are handling it correctly (see Chapter 6). It will not be practical to purify some reagents, for example alkyllithiums, and in this case the quality should be checked by titration where possible. It is unwise to purchase a reagent (from any source) and to take for granted the quoted molarity and purity; even the most reputable suppliers are fallible and occasionally make mistakes.

Once you are certain that the problem does not lie with the starting material or reagents, check the solvent. Many reactions will not work if the solvent is not anhydrous; methods for obtaining anhydrous solvents are given in Chapter 5. Tetrahydrofuran (THF) is very commonly used, and is usually dried by distillation off sodium/benzophenone; however it is possible collect *wet* THF from a bright blue distillation pot (which means that the solvent is dry in this part of the still). This problem is encountered when a still head is used to collect the solvent, but insufficient time has been allowed at reflux before collection is commenced. One way to check this is to repeat the reaction, but take a sample of the THF used (before introducing it into the reaction flask) and add a little sodium hydride (dispersion in oil) (*care*). If immediate hydrogen evolution is observed, the solvent is wet. The remedy is to allow more time at reflux before collecting the solvent, if this does not work the drying agents might need recharging.

Another source of water could be the inert gas which you are using; make sure that the drying agent (if used) used is still working renew it if necessary. Failing this, try the reaction under argon, which usually contains much less moisture than nitrogen. Check the manifold and renew any suspect tubing.

If the starting material remains unreacted after the attempted reaction, and if starting material, reagents, solvent, and inert gas are all as they should be, consider the possibility that the reaction might be reversible. It might be that you are actually under equilibrating conditions and that the reaction is under "thermodynamic control" and the equilibrium might lie in favour of the starting materials. Conditions and reagents might need to be changed to avoid this, you might require conditions of "kinetic control". Only then consider that it might be you! You might be inadvertently carrying out the reaction in such a way that it will not work. For example, is the temperature correct? Is the *concentration* of reactant and reagent correct? Is too much or too little time being allowed at a particular stage? There are many possibilities. To test this it is advisable to carry out the same reaction but use a substrate which is known from the literature to react properly. If this is successful, and your desired reaction is not then you might have found a reaction which does not work on your substrate, and alternative conditions (different metal ions, different Lewis acid etc.) might be required. Above all, if the reaction is an important one, do not give up; perhaps you could take the opportunity to develop a new reagent or procedure which will work!

If the starting material is decomposed under the reaction conditions, then consider carefully any possible "chemical" reasons which might be responsible. For example, basic conditions might be causing deprotonation and/or elimination at an undesired site or sites in the molecule, and acidic conditions might be responsible for unexpected cyclisation or protecting group loss. There are often several possible sources of instability towards a particular set of reaction conditions, and it will be necessary to examine all possibilities.

It is likely that the reaction will be closely related to a known process, and if this is so then be critical in your comparison of reaction conditions and reactants. For example if an enolate alkylation is being attempted, but appears not to work at a temperature below which the enolate decomposes, then consider the reactivity of the alkylating reagent. Is it really as reactive as in the example which you are following? If not, then the appropriate change will need to be made.

Decomposition of your starting material might be occurring at several stages in the reaction. This can often be ascertained by careful tlc of the reaction mixture at various stages. Take the hypothetical enolate reaction referred to above. Decomposition might be occurring on addition of base, before addition of the alkylating agent. Analysis by tlc at this stage (before addition of alkylating agent) will reveal decomposition, and tlc after addition will also indicate decomposition at this stage. If all looks fine, and a product appears to have been formed (by direct tlc of the reaction mixture) then the product might be decomposing on work up. This is easily checked by quenching samples of the reaction mixture into solutions of varying pH and buffer solutions, followed by tlc analysis.

If all else fails, and you are attempting to follow published a procedure, or to apply a published procedure to a "new" system, then it might be worth contacting the senior author on the publication which you are following. Many such authors are willing to provide further information and advice. If such help is forthcoming and allows you to succeed, then be sure to let the supplier of such information know, and send a copy of your manuscript should you wish to publish your results. Be sure to acknowledge this help in such publications. This simple courtesy will avoid any misunderstandings and ensure good relations in the future!

The single most important piece of advice in "trouble-shooting" a reaction is to be certain that you identify the problem, *before you try to remedy it*. This might seem to be obvious, but it is surprising how often this advice is not followed. Try hard to keep an open mind, to analyse the problem critically and logically, and to gather evidence to support your identification of the problem. When you have a clear idea what the problem actually is, you can begin to think of potential solutions which are likely to have the desired effect.

CHAPTER 17

The Chemical Literature

17.1 The structure of the chemical literature

17.1.1 Introduction

Consulting the literature is an essential element of chemical research. Whether you want to confirm the identity of your latest product, or check the feasibility of an exciting new idea, it would be both unscientific and counter-productive not to conduct a thorough literature search. Moreover, it is vital to the success of your work to keep abreast of developments in organic chemistry in general, and in your area in particular. The problem is that finding chemical information and keeping in touch with current developments are difficult and time consuming tasks.

We are the beneficiaries of almost one and a half centuries of research in organic chemistry. The accumulated output of that effort is an enormous body of data collected in a vast literature. It is estimated that well over half a million articles (papers, patents, books, etc.) are published each year and the volume of publications will probably continue to rise. Searching such a huge body of work is a formidable problem but it must be emphasized that the time spent reading the literature is often more than repaid by the experimental time saved as a result.

This chapter is intended as a practical guide to efficient searching of the chemical literature; the main part is devoted to a description of the most important access routes to the primary literature, and a discussion of methods of tackling some common types of literature search. The chapter concludes with a section on methods of keeping in touch with the current literature. The reader is referred to a number of recent texts for more detailed information.[1-4]

17.1.2 The structure of the literature

Almost all chemical information is originally published in research journals, in patents, and in theses. These sources are called the primary literature and the goal of most literature searches is to find the original reports containing the required information. There are thousands of journals which publish papers on chemistry but in practice the great majority of papers which are of interest to the organic research chemist appear in just a hundred or so of these. This is still a dauntingly large body of information but there are several routes by which it can be searched and specific items of information located.

An important route, and one which is rapid and easy, is to tap the chemical knowledge of your colleagues and supervisors. Many of the people working around you are likely to be experts in their own fields. Another route is to use the secondary literature. This comprises review articles and books, in which the original literature has been organized and summarized, and reference books in which particular kinds of data have been collected together. Of course, finding the appropriate review or handbook is a problem in itself. A third route is *via* indexes which give the literature references for all of the information on a given compound, or procedure, or author, etc. Prominent among these is Chemical Abstracts, which covers over 12,000 different sources and contains short summaries of just about every paper published on a chemical topic, as well as comprehensive indexes to these abstracts, and hence to the original papers. Finally there are a wide range of computer databases, which offer unprecedented speed, reliability, and flexibility, but usually at a price!

In order to aid discussion of these information sources we have split them into two categories; paper-based information sources (journals, books, series, etc.) and electronic sources (computer databases), because although the type of information available from each source may be the same, the coverage and methods for searching them often differ significantly.

1.   G. Wiggins, *Chemical Information Sources*, McGraw-Hill, New York, 1991.

2.   H. Schultz, *From CA to CAS ON-LINE*, VCH Verlagsgesellschaft, Weinheim, 1988.

3.   Y. Wolman, *Chemical Information, A Practical Guide to Utilization*, 2nd ed., Wiley, Chichester, 1988.

4.   R.E. Maizell, *How to Find Chemical Information*, 2nd ed., Wiley, Chichester, 1987.

17.2 Some important paper-based sources of chemical information

This section contains a description of structure, strengths, and weaknesses of three of the most important paper-based tools for locating information in the primary literature: Chemical Abstracts, Beilstein, and the Science Citation Index. It is followed by a complementary section on how to carry out some specific kinds of searches.

17.2.1 Chemical Abstracts

Chemical Abstracts (CA) consists of two main parts, abstracts of every paper containing new chemical information, and indexes which provide access to the abstracts and thence to the original literature. It is published weekly and each issue contains a keyword index and an author index. The weekly issues are collected in volumes covering a six month period (one year, prior to 1962) and each volume contains author, chemical substance, formula, general subject, patent, and ring system indexes. Every five years (ten years, prior to 1957) the indexes for the ten volumes are combined to give *Collective Indexes*. These indexes are the single most important and comprehensive information tool available to the chemist.

A search of Chemical Abstracts should begin with the appropriate index of the most recent volume and should progress backwards through the other volumes until the beginning of the period covered by the most recent Collective Index (currently the 12th - 1987-1991), at which stage the Collective Indexes are used to search the literature back to 1907. The most useful indexes are the chemical substance, formula, and general subject indexes and their use is described more fully in Section 17.3. Consulting the indexes will, in the first instance, lead to references to the abstracts, not directly to the literature. The references to the abstracts take the following form: **90**:*108753h* where **90** is the CA volume number and the abstract number is *108753*. Since 1967 abstracts have been numbered sequentially in each volume. Prior to that the references were of the form **46**:*13761a* where **46** is the volume number, *13761* is the column number, and the letter *a* indicates that the abstract is at the top of the column. Earlier still, a numerical superscript was used to indicate the position of the abstract in the column. The letters *B*, *P,* or *R* before an abstract number indicate that the original work is a book, patent, or review, respectively.

The abstracts contain full bibliographic details of the original paper, and a summary of the principal new findings reported in the paper. A glance at the abstract will tell you if the original journal is likely to be accessible, what language the paper is in, and most importantly it will give an indication of whether the paper really does contain the information you require. Remember that many of the compounds described in the original paper will not be mentioned in the abstract but will be contained in the indexes. If the abstract looks promising all that remains is to locate the journal and consult the paper.

An *Index Guide* is published every eighteen months and contains invaluable information on the use of the indexes and the system of nomenclature used in CA. It is essential reading for serious users of Chemical Abstracts. Finally, the *Ring Systems Handbook* and its predecessors, the *Ring Index* and the *Parent Compound Handbook*, contain information on ring and cage systems and gives the names under which ring systems can be found in the Chemical Substance Index.

The great strengths of CA are that it provides comprehensive literature coverage, and it has extensive indexes. However, the coverage of the literature in the early years was not so rigorous, and Beilstein provides more thorough coverage of the pre-1949 literature.

17.2.2 Beilstein

Beilstein's Handbuch der Organischen Chemie, or Beilstein for short, is a huge (> 300 volumes) reference work which contains physical and chemical data for over one and a half million compounds. The compounds are organized according to a unique classification system and each volume contains a subject (actually compound) index and a formula index. Comprehensive literature coverage is attempted, so Beilstein contains essentially all the compounds of a given class, which were prepared from the beginning of organic chemistry to the date of publication of the most recent volume covering that class of compounds. A considerable amount of critically reviewed information, with references to the primary literature, is provided for each compound. This data includes the molecular structure, natural occurrence, methods of preparation, physical properties including references to papers containing spectral data, and chemical properties.

The Handbook consists of the original series of 29 volumes (the *Hauptwerk, H*) covers the literature up to 1909, and is divided as follows: volumes 1-4, acyclic compounds; volumes 5-16, carbocyclic compounds; volumes 17-27, heterocyclic compounds; volume 28, General Subject Index; and volume 29, General Formula Index. In addition there are a four complete supplementary series (*Erganzungswerk, EI, EII*, etc.) covering the literature up to 1959. Work on the supplements is continuous, and some progress has been made in bringing the coverage up to 1979 (the fifth supplement). Cumulative indexes are only available for the pre-1930 literature. It is relatively easy to find data for any compound reported prior to 1930 by checking the cumulative formula index (Volume 29 in three sub volumes) which will give the volume and page numbers for the entries in H, EI and EII. Compounds of the same class will appear in the same volume of each series (although some of the volumes are now divided into several sub volumes). Thus if a compound is located using the cumulative index it is a simple matter to locate references in the later series, using the volume indexes for the same volume or using the *Beilstein System Number*. If a compound is not contained in the cumulative index it is best to try to identify which volume it should be contained in, using the classification system. A booklet explaining how to use the Beilstein system is available from the publishers.[5]

Beilstein is the best and most comprehensive source of data for organic compounds prepared before 1930 and it is not particularly difficult to use. Its major weakness is the lack of data and cumulative indexes for more modern work.

17.2.3 Science Citation Index

Science Citation Index (SCI) is a combination of three indexes which provide coverage of all the important publications in the physical sciences. SCI is published every two months and is cumulated annually. There are cumulative indexes covering the period 1945-1979. It includes coverage of all of the major chemistry journals.
1.   The *Source Index* lists the bibliographic details for the publications for each author/organization.

5.   *How to Use Beilstein*, Beilstein Institute, Springer-Verlag, Berlin, 1978.

2. The *Permuterm Index* is based on combinations of keywords in the titles of the articles published in the journals which are covered by SCI. For example under the keyword 'epoxide' will be a list of other keywords, such as 'stereoselective', which occur in association with 'epoxide'. For each pairing there is a list of authors names, and looking these up in the *Source Index* will lead to the references for the original work.

3. The *Citation Index* is a unique feature which allows you to search the literature *forward* in time. The index entries are the names of the first authors of each paper which was cited in any paper published during the period covered by that issue. Its use is best illustrated with the aid of an example. Suppose that a researcher found a paper published in 1980, by S. Smith *et al.*, which contained some very interesting results, and he wanted to know if any further relevant work had appeared. He would consult the annual indexes of the SCI from 1980 onwards. Each index contains an alphabetic list of *first* authors names and under S. Smith's name would be a chronological list of his/her publications. Under the entry for the paper of interest to our researcher there would be a list of papers, published during the period covered by that index, which cited it. It is reasonable to assume that any workers who followed up on the results in the Smith paper would have cited it in their own publications. Thus the list of papers which cited the original should include most of the work subsequently carried out in that area. Hence if you find an important paper you can use SCI to get a list of all the papers which subsequently referred to it. The drawback is that many of the references you find will not be relevant to your interest and there is no way of knowing which are relevant except by consulting the *Source Index*, which gives the titles of the papers, or by consulting the papers themselves.

The citation index is an extremely useful tool and we recommend that you carry out a search for every key paper you come across. You can also use it to find out who is referring to your own work.

17.3 Some important electronic-based sources of chemical information

The development of computer databases over the last decade has brought about a revolution in the way in which we search and store chemical information, and many of the paper-based information sources are now available in electronic form. The advantages of so-called 'on-line searching' include much greater speed, greater accuracy, and greater reliability. Some computer databases include material which cannot otherwise be searched directly, e.g. the full text of many major journals and reference books, and they generally contain more information than is accessible at the majority of institutional libraries. The principal advantage is much greater flexibility and power in carrying out searches. For example, it is possible to combine a number of keyword searches in one using logical operators (oxidation AND (alcohol OR aldehyde)). Even more importantly it is possible to search structures or substructures *via* graphical input, searches which were almost impossible using printed indexes.

There are disadvantages too. Some databases do not include all of the material available in conventional form; for example, much of the early Chemical Abstracts is not available on-line. Another problem is that the software for on-line searching is relatively complex, so it is more difficult to learn and, as a result, on-line searches by inexperienced users can be unreliable. The increasing use of personal computers and graphical input and output have made database searching much easier but a good understanding of the software is still essential. Finally the cost of hardware, software, consumables, and the searches themselves, can be considerable.

Below we briefly outline details of some of the more important chemical databases available. All these databases are commercial and require registration and payment of an annual fee to use, although in the UK the Cambridge Structural Database and REACCS are available free to the academic community as part of the Chemical Databank Service (CDS) based at Daresbury, Warrington. In general access to such databases will vary from institute to institute and as there are a wide range of different chemistry databases potentially accessible; you should ask at your library for a list of those available to you. It is not possible to describe the operation of all the databases here, but you should be able to get help from your librarians or your supervisor and learn how these very powerful tools can help you.

17.3.1 Beilstein

Although incomplete, the electronic version of Beilstein is now available by subscription and constitutes a very powerful means of searching this important database. Currently this database contains information on about 5 million compounds and is constantly being updated. Searches can be performed using keyword or structural input and will provide the same type of information as found in the paper-based version of this database.

17.3.2 CAS ON-LINE

This is the electronic form of *Chemical Abstracts* and consequently constitutes the largest database available that relates to synthetic organic chemistry. The database allows the searching of Chemical Abstracts back to 1957 (with a limited number of records prior to this time), although complete retrieval of bibliographical and abstract information is only possible back to 1970. It features powerful keyword and graphical search facilities and is an essential resource for full-time organic chemists.

17.3.3 CASREACT

This is a reaction database based on *Chemical Abstracts*, and contains information on a huge number of chemical reactions taken from publications during the period 1985-present. It has powerful graphic searching facilities and displays detailed reaction schemes, bibliographic information and abstracts. It is an important source of information for anyone about to attempt a chemical transformation with which they are unfamiliar.

17.3.4 SCI

This is the electronic version of the Science Citation Index and can be searched by a variety of text inputs including keywords, author's name, institute name, and citation. Currently it covers the time period 1981-present and is a very rapid alternative to manually searching this invaluable publication.

17.3.5 Chemical Journals On-line (CJO)

Chemical Journals On-line is a database that contains the full text of many chemical journals. This database does not give complete coverage of all chemistry journals and does not cover issues published before 1982, but it does allow rapid keyword searching of the full text of those included, and short portions of the articles along with bibliographical data can be displayed. Consequently it is a very powerful means of searching the journals that are included and is particularly useful where the original paper versions of the journals are unavailable.

17.3.6 REACCS (Molecular Design Ltd.)

REACCS is a large commercial reaction database marketed by Molecular Design Ltd. It contains reaction data obtained from a variety of sources including the *Journal of Synthetic Methods, Theilheimer's Synthetic Methods of Organic Chemistry*, and *Organic Syntheses*. It currently contains about 250,000 reaction entries and has powerful graphic searching facilities. It is similar to *CASREACT* in the information it provides and although this database is not as comprehensive, because the majority of the contents have been selected for their usefulness to synthetic chemists it is an excellent alternative.

17.3.7 Cambridge Structural Database

This database contains bibliographic, and 3D structural results from crystallographic analyses of organics, organometallics and metal complexes. Both X-ray and neutron diffraction studies are included. The database covers structures up to about 500 atoms (including hydrogens) and as such is an invaluable source of structural information. Currently there are about 127,000 entries which are increasing at the rate of about 12,000 a year. It can be searched by both keyword and graphical input.

17.3.8 The World Wide Web

Although not strictly a database, the World Wide Web (WWW) is rapidly becoming an important source of all types of information, including chemical information. The WWW is freely accessible to anyone in an academic institute around the world who has a suitably networked computer, and if you are unsure about access you should consult your local computing centre. It is also possible for non-academics to access the WWW either directly or *via* a variety of commercial computer bulletin boards. Sites on the WWW are linked by a global network often referred to as *the internet*, and information at these sites can be accessed using a range of free or inexpensive software packages such as the excellent *Mosaic* (available for a variety of computing platforms, from the National Centre for Supercomputing Applications, University of Illinois, USA). This software provides a user-friendly interface and allows you to navigate through the WWW simply by pointing a cursor and clicking the mouse-button. In addition you can connect to different sites specifying a code known as a URL (uniform resource locator). URL's in the format used by Mosaic for the sites mentioned in this section are included in brackets after the name of the site. These codes will work with most WWW software although in some cases the format may change slightly. The chemical information available on the WWW is expanding rapidly and includes a variety of specialized databases such as the Brookhaven Protein Databank, best accessed *via* Molecules R US (http://www.nih.gov/htbin/pdb), which includes structural information on proteins and nucleic acids. If you have access to the WWW the best way to find out the range of chemical information available is to consult the chempointers index (http://www.chem.ucla.edu/chempointers.html) which will direct you to the majority of chemistry sites available. The main advantage of searching chemical information *via* the WWW is that most of the sites allow free use of the databases, although this situation may change in the future as the commercial potential of this network is realised. The disadvantages are that at present the range of databases available is limited so you may not be able to find the information required, and that access to the WWW can become very slow at peak times during the day.

17.4 How to find chemical information

17.4.1 How to do searches

The following are some basic rules for guidance in searching the literature.
1.   Clearly define the goals of your search.

2. Discuss the problem with your colleagues and supervisors; they may have some valuable expertise.

3. Determine which information sources are available to you and which to use.

4. Start with the current literature and work backwards, the recent literature will contain references to earlier work.

5. When you find a key paper, check it carefully for references to relevant earlier work, and also work forward in time by carrying out a Science Citation Index search.

6. Keep a complete record of your search, noting all the sources you used and the information you obtained (e.g. lists of CA abstract numbers and their contents). This is invaluable if you have to carry out related searches later. It is advisable to keep a separate notebook for recording your literature searches, the information you accumulate will build into a very useful resource. If you perform electronic searches it is usually possible to download the search results onto a computer disk in a format that you can refer to again later if necessary

17.4.2 *How to find information on specific compounds*

The chief sources of data for particular compounds are Chemical Abstracts and the numerous reference handbooks, including Beilstein. Information on relatively simple compounds can often be obtained from handbooks such as the following:

1. The catalogues of major chemicals suppliers.

2. *CRC Handbook of Chemistry and Physics*, R.C. Weast Ed., CRC Press.
    CRC Handbook for Organic Compound Identification.
    CRC Handbook of Data on Organic Compounds.
    These sources contain physical and chemical data for a large number of organic compounds.

3. *The Dictionary of Organic Compounds,* 5th ed. Chapman and Hall.
    The Dictionary of Organometallic Compounds.
    These multi-volumed works quote physical properties and references to the preparation and properties of about 75,000 compounds and many derivatives. Both have name and formula indexes, kept up to date by the publication of annual supplements. They are also available on-line.

4. Several extensive collections of spectral data are available. The most extensive of these are produced by Sadtler Research Laboratories and cover ir, Raman, uv, ^{1}H nmr, ^{13}C nmr, and mass spectra. The Aldrich Chemical Company has produced three excellent collections of data: ir, FT-ir, and ^{1}H nmr.

For data on more complex structures it is usually necessary to turn to Beilstein or Chemical Abstracts. Beilstein is in fact a giant handbook containing data for *all* organic compounds published in the timespan of the volumes which are available. Its use is described in Section 17.2.2. The best way to find specific compounds in Chemical Abstracts is to start with the *Formula Index*. The formulae are listed in order of increasing number of carbons, then increasing number of hydrogens, and then increasing numbers of the other elements in alphabetical order. Under each formula is a list of names, and for each substance there is a list of abstract numbers. It is best to scan through the list of names to try to identify the compound you want and then, armed with the correct CA name, use the *Chemical Substance Index*. The *Chemical Substance Index* has the advantage that, for each substance, it gives a list of keywords (isolation, preparation, etc.) followed by the abstract numbers. The keyword list makes it much easier to identify which abstracts are most likely to contain the information you require. With a little experience you may prefer to use the *Chemical Substance Index* directly but the complexities of nomenclature and indexing preclude any further discussion here. Note that prior to the 9th Collective Index, the *Chemical Substance* and *General Subject Indexes* were combined in the *Subject Index*.

17.4.3 *How to find information on classes of compounds*

Finding information about a broader area, such as a class of compounds is usually more difficult than finding data about a specific compound. It is usually best to begin by consulting books on the area, and then progress to more specialized monographs and reviews, before consulting the primary literature.

Good starting places include *Comprehensive Organic Chemistry, Comprehensive Organometallic Chemistry* and *Comprehensive Heterocyclic Chemistry* (Pergamon Press), which are multi-volume texts giving a detailed overviews of the title areas. A much more detailed treatment of many common classes of compounds is contained in the series *The Chemistry of the Functional Groups* edited by S. Patai (Wiley). This excellent series consists of over 30 volumes (in nearly 60 parts) each of which contains reviews on all aspects of the chemistry of one particular functional group. Other multivolume series include *The Chemistry of Heterocyclic Compounds - A Series of Monographs* (Wiley) and *Rodd's Chemistry of Carbon Compounds* (Elsevier). In addition there are many review series devoted to the chemistry of particular classes of compounds, including the *Specialist Periodical Reports* published by the Royal Society of Chemistry. See Section 17.4.4 for a list of sources of information on synthetic methods for families of compounds.

Finding individual books or reviews is more difficult. A good starting point is your library, glance along the shelves and consult the catalogue. A more systematic method is to use *Index of Reviews in Organic Chemistry* (Royal Society of Chemistry) or *Index to Scientific Reviews* (Institute for Scientific Information) to locate books and reviews.

Manual searching of *Chemical Abstracts* is not a good method for tackling this kind of search because the indexing policy means that very few articles will be cited under a general heading such as aldehydes. *CAS ON-LINE* will usually give much better results because a wider range of subject terms can be used in the search. If you are looking for information on a highly specific structure type, a substructure search of *CAS ON-LINE* should give essentially 100% recovery of the relevant recent references.

17.4.4 How to find information on synthetic methods

The reference books on the chemistry of classes of compounds which are listed in Section 17.4.3 are good starting points in this case too. Additionally there are several major works devoted specifically to synthetic methods. Most notable are the *Encyclopaedia of Reagents for Organic Synthesis* (Wiley) an eight volume work which contains short reviews on almost all the common reagents used in organic synthesis, and *Reagents for Organic Synthesis* (L.F. Fieser and M. Fieser, Wiley) which is a continually updated multi-volume series of similar format. These two works constitute the best source of information on the preparation, purification, and use, of synthetic reagents. An inexpensive alternative is the single volume work *Comprehensive Organic Transformations* (R.C.

Larock, VCH Publishers) which outlines a large range of functional group transformations and is an excellent starting place when searching for such information.

Other useful texts include *Comprehensive Organic Synthesis* (Pergamon), a nine volume overview of synthetic methods, *Compendium of Organic Synthetic Methods* (Wiley), *Organic Reactions* (Wiley), *Advanced Organic Chemistry, Reactions, Mechanisms, and Structures* by J. March (Wiley), and *Organic Syntheses* (Wiley), a compilation of carefully checked procedures with full experimental details which is an excellent source of representative synthetic procedures.

Two excellent series provide annual coverage of developments in synthetic chemistry. Theilheimer's *Synthetic Methods of Organic Chemistry* provides thorough coverage of the synthetic literature and is organized according to a unique (and easily learned) system of classifying the transformations taking place. Theilheimer is also available on-line as part of *REACCS* (see Section 17.3.6). *Annual Reports in Organic Synthesis* (Academic Press) is a well organized collection of representative synthetic transformations and it is up to date and very easy to use.

In addition a number of electronic databases are particularly useful for searching synthetic methods most notable amongst these being *CASREACT* and *REACCS* both of which allow access to a huge number of transformations (see Section 17.3).

17.5 Current awareness

Keeping in touch with the current literature is a difficult and time consuming exercise but it is vital to your development as a chemist, and to the success of your research. You should aim to read through at least 6-12 of the most important journals in your field and the best way of doing this is to set aside a specific period each week for reading the periodicals. You should also scan the review journals (*Angewandte Chemie, International Edition in English, Chemical Society Reviews, Chemical Reviews, Synthesis,* and *Tetrahedron*) and magazines such as *Chemical and Engineering News, Chemistry and Industry* and *Chemistry in Britain.*

As if this is not enough, there is still the problem of how to cover the hundreds of other chemistry periodicals. The only practical method of doing this is to use compilations of abstracts. You could read Chemical

Abstracts itself but this is too large and a much better choice is one or more of the titles in the *CA Selects* series, which only contain abstracts relating to a particular area. However, wider literature coverage is essential and a good approach is to read one of the periodicals which abstracts new compounds and reactions. *Methods of Organic Synthesis* (Royal Society of Chemistry) and the more comprehensive *Index Chemicus* (Institute of Scientific Information) are good examples of this genre.

All of this effort will be wasted if you do not keep good records of what you have read. Building your own computer database is an increasingly practical way of doing this but for now the simplest method is to use a card file. Make a record of each important paper on an index card. Include the bibliographic details and an abstract of the key results in the paper. The pile of cards is not of much use unless it is properly filed. Many filing systems, each with its own strengths and weaknesses, can be conceived but one possibility deserves special mention, at least as a starting point. *Annual Reports in Organic Synthesis* (Section 17.4.4) is essentially a compilation of index cards in book form, and the systematic way in which the material is organized could serve as a useful model for your system. As time progresses you can modify this system to adapt it to your own interests and requirements.

Appendix 1 : Properties of common solvents

Solvent	B.p. (°C)	M.p. (°C)	ε	Density	δ¹H (Ref. to TMS) (300MHz, CDCl₃)	δ¹³C (Ref. to TMS) (300MHz, CDCl₃)	Preliminary drying	Rigorous drying	TLV (ppm)
Acetic acid	118	17	6.19	1.049	2.08 s, 10-13 br s var	20.7, 177.6	Acetic anhydride	Acetic anhydride	10
Acetone	56	-94	20.7	0.790	2.13 s	30.7, 206.5	3A sieve	3A sieve	1000
Acetonitrile	82	-46	36.2	0.777	1.97 s	1.7, 116.2	Potassium carbonate	Phosphorus pentoxide	40
Benzene	80	5.5	2.28	0.879	7.37 s	128.3	Not necessary	Calcium hydride	10
t-Butanol	82	25	3.49	0.850	1.24 s, 1.35 br s var	31.2, 69.2	Calcium hydride	Calcium hydride	10
Carbon tetrachloride	76	-23	2.23	1.460	-	96.2	Alumina	Phosphorus pentoxide	
Chlorobenzene	132	-46	5.62	1.106	7.28 br m	126, 129, 130, 134	Not necessary	Calcium hydride	75
Chloroform	62	-63	4.70	1.480	7.24 s	77.0	Alumina	Phosphorus pentoxide	10
Dichloroethane	83	-35	10.4	1.235	3.71 s	44.4	Not necessary	Calcium hydride	10
Dichloromethane	40	-97	8.9	1.325	5.28 s	53.4	Not necessary	Calcium hydride	100
Diethyl ether	35	-116	4.34	0.714	1.18 t, 3.45 q	15.2, 65.8	Calcium chloride; Na	Sodium/benzophenone	400
Dimethoxyethane	83	-58		0.850	3.36 s, 3.51 s	59.0, 71.8	Calcium chloride; Na	Sodium/benzophenone	
Dimethylformamide	152	-61	36.7	0.945	2.81 s, 2.89 s, 7.94 s	31.4, 36.4, 162.4	Calcium hydride	Phosphous pentoxide	10
Dimethyl sulphoxide	189	18	49.0	1.101	2.62 s	40.6	Distillation	4A sieve	
Dioxan	102	12	2.21	1.034	3.66 s	67.0	Calcium chloride: Na	Sodium/benzophenone	50
Ethanol	78	-114	24.3	0.785	1.18 t, 2.05 br s, 3.65q	18.3, 58.2	Magnesium	3A sieve	1000
Ethyl acetate	77	-84	6.02	0.900	1.22 t, 2.01 s, 4.09 q	14.1, 20.9, 60.3, 171.0	Potassium carbonate	4A sieve	400
HMPA	235	7		1.030	2.59 d	36.8	Calcium hydride	Calcium hydride	
Methanol	64	-97	32.6	0.791	1.94 br s var, 3.42 s	50.5	Magnesium	3A sieve	200
Nitromethane	101	-28	38.6	1.137	4.33 s	62.4	Calcium chloride	4A sieve	100
Pyridine	116	-42	12.3	0.982	7.24 m, 7.63 m, 8.58 m	123.6, 135.8, 149.8	Calcium hydride	Calcium hydride	5
Tetrahydrofuran	66	-108	18.5	0.805	1.82 m, 3.72 m	25.6, 67.9	Calcium chloride; Na	Sodium/benzophenone	200
Toluene	111	-95	2.38	0.867	2.32 s, 7.17 br s	21,125,128,129,138	Not necessary	Calcium hydride	100
Water	100	0	78.5	1.000					

Appendix 2 : Properties of common gases

Gas	Mol. weight	B.p.^{a}	M.p.^{b}	Density of gas^c	Density of liquid^d	Hazardous properties^e	TLV^f
Acetylene	26.04	-84	-78	1.109	0.68	As, Fl, Note g	25
Ammonia	17.03	-33	-78	0.71		To, Co, Fl	
Argon	39.944	-189	-186	1.66	1.4	As	
Boron trichloride	117.19	12.5	-107	4.85		To, Co	
Boron trifluoride	67.81	-100	-127	3	1.59	To, Co	1
Carbon dioxide	44.01		-78	1.83		Co	5000
Carbon monoxide	28.01	-191	-207	1.16	0.79	To, Fl	50
Chlorine	70.914	-34	-101	2.97	1.56	To, Co	1
Ethylene	28.05	-104	-170	1.17	0.57	Fl	
Ethylene oxide	44.05	10.7	-112	1.82	0.88	To, Fl	5
Fluorine	37.997	-188	-220	1.57	1.5	Fl, Co	1
Helium	4.0026	-269	-272	0.17	0.12	As	
Hydrogen	2.016	-253		0.08	0.07	Fl	
Hydrogen bromide	80.917	-67	-87	3.34	2.16	To, Co	3
Hydrogen chloride	36.461	-85	-114	1.52	1.19	To, Co	5
Hydrogen fluoride	20.006	19.5	-83	0.94	1.0	To, Co	3
Hydrogen sulphide	34.08	-60	-85	1.43		To, Co, Fl	10
Isobutylene	56.11	-6.9	-140	2.39	0.63	Fl	
Methanethiol	48.107	6	-121	2.14	0.89	To, Fl	0.5
Nitric oxide (NO)	30.006	-152	-164	1.24	1.27	To	25
Nitrogen	28.0134	-196		1.25	0.8	As	
Nitrogen dioxide (NO$_2$)	46.0055	21	-9.3	3.3	1.45	To, Co	3
Oxygen	32.0	-183	-218	1.33	1.14	Ox	
Phosgene	98.92	8.2	-128	4.1	1.41	To, Co	0.1
Sulphur dioxide	64.063	-10	-75	2.70	1.46	To, Co	2

Notes

a. °C at 1atm
b. °C
c. g/l at 20°C at 1atm
d. g/ml at b.p.
e. As = Asphyxiant
   Co = Corrosive
   Fl = Flammable
   Ox = Oxidising
   To = Toxic
f. ppm
g. Potentially explosive when pressurised

Appendix 3 : Approximate* pK$_a$ values for some common deprotonations cf. some common bases

Reagent to be deprotonated	pKa	Bases	pKa (of BH)
ArSH	7		
RCH$_2$NO$_2$	9		
RCOCH$_2$CN	9		
RCOCH$_2$COR	9		
RSH	10	Et$_3$N	11
RCOCH$_2$CO2R	11	Et$_2$NH	11
RCO$_2$CH$_2$CO$_2$R	13		
Cyclopentadiene	15	Na$^+$ $^-$OH	16
RCH$_2$CHO	17	Na$^+$ $^-$OEt	18
RC≡CH	25	K$^+$ $^-$O^{t}Bu	19
CH$_3$COCH$_3$	20	Li$^+$ N(TMS)$_2$	26
CH$_3$CO$_2$Et	25	Na$^+$ $^-$H	35
CH$_3$CN	25	Na$^+$ $^-$NH$_2$	35
PhH	41	Li$^+$ $^-$N^{i}Pr$_2$	36
H$_2$C=CHCH$_3$	43	Li$^+$ TMP	37
H$_2$C=CH$_2$	44	Li$^+$ $^-$^{n}Bu	50
		Li$^+$ $^-$^{t}Bu	>50

*These figures are **very** approximate. A pK$_a$ difference of >4 between base and reagent will cause complete deprotonation.

Appendix 4 : Lewis acids

Lewis acid	Compatible solvents	Comments
Aluminium trichloride	Hydrocarbons, halogenated	Strong, widely used in Friedel-Crafts
Boron tribromide	Hydrocarbons, halogenated	Strong, used to cleave ethers, acetals
Boron trichloride	Hydrocarbons, halogenated	Strong, used to cleave ethers, acetals
Boron trifluoride etherate	Many solvents	Moderate, very versatile
Diethylaluminium chloride	Hydrocarbons, halogenated	Moderate, useful for proton-sensitive reactions
Ethylaluminium dichloride	Hydrocarbons, halogenated	Moderate, useful for proton-sensitive reactions
Ferric chloride	Hydrocarbons, halogenated	Moderate
Mercuric chloride	Many solvents	Weak, useful for cleaving C-S bonds
Lanthanide shift reagents	Many solvents	Weak, useful for reactions involving sensitive dienes, ethers
Magnesium bromide	Hydrocarbons, halogenated	Moderate
Magnesium chloride.etherate	Hydrocarbons, halogenated, ethers	Moderate
Silver chloride	Many solvents	Moderate, used to generate carbonium ions
Silver triflate	Many solvents	Moderate, used to generate carbonium ions
Stannic chloride	Hydrocarbons, halogenated	Strong, very versatile
Titanium tetrachloride	Hydrocarbons, halogenated	Strong, very versatile
Trimethylsilyl iodide	Hydrocarbons, halogenated, MeCN	Strong, used to cleave ethers, acetals, esters
Trimethylsilyl triflate	Hydrocarbons, halogenated	Strong, used with silylated reagents
Zinc bromide	Hydrocarbons, halogenated	Moderate, versatile
Zinc chloride	Hydrocarbons, halogenated, ethers	Moderate, versatile

Appendix 5 :   Common reducing reagents

1. Hydride reducing agents

Reagent	Typical solvents	Temperature (°C)	Functional groups reduced
LiBH$_4$ (lithium borohydride)	Tetrahydrofuran	0 to RT	ester → alcohol ketone → alcohol aldehyde → alcohol
Li[Et$_3$BH] (Superhydride) (lithium triethylborohydride)	Tetrahydrofuran	-78 to RT	ester → alcohol ketone → alcohol aldehyde → alcohol alkyl halide → alkane epoxide → alcohol
Li[^{s}Bu$_3$BH] (L-Selectride) (lithium tri-*sec*-butylborohydride)	Tetrahydrofuran	-78 to RT	ketone → alcohol aldehyde → alcohol alkyl halide → alkane
NaBH$_4$ (sodium borohydride)	Alcohols, ethers	0 to RT	ketone → alcohol aldehyde → alcohol
Na[BH$_3$CN] (sodium cyanoborohydride)	Alcohols, water, DMSO	0 to RT	ketone → alcohol aldehyde → alcohol alkyl halide → alkane imine → amine
Na[BH(OAc)$_3$] (sodium triacetoxyborohydride)	Acetic acid	0 to RT	ketone → alcohol aldehyde → alcohol

Appendix 5 cont'd

Reagent	Typical solvents	Temperature (°C)	Functional groups reduced
$Zn(BH_4)_2$ (zinc borohydride)	Ethers	0 to RT	ketone → alcohol aldehyde → alcohol
$^nBu_4NBH_4$ (tetra-n-butylammonium borohydride)	Dichloromethane, ethers	0 to RT	ketone → alcohol aldehyde → alcohol
$LiAlH_4$ (lithium aluminium hydride)	Ethers	-78 to RT	ester → alcohol ketone → alcohol aldehyde → alcohol alkyl halide → alkane acetylene → trans alkene epoxide → alcohol imine → amine amide → amine
$Li[(^tBuO)_3AlH]$ (lithium tri-tert-butoxyaluminium hydride)	Tetrahydrofuran	-78 RT	acid chloride → aldehyde ketone → alcohol aldehyde → alcohol
$Na[AlH_2(OCH_2CH_2OCH_3)_2]$ (RED-AL) (sodium bis-[2-methoxyethoxy]-aluminium hydride)	Ethers, toluene	-78 to RT	ester → alcohol ketone → alcohol aldehyde → alcohol alkyl halide → alkane epoxide → alcohol α,β-unsaturated enone → allylic alcohol

Appendix 5 cont'd

Reagent	Solvent	Temperature	Reductions
B₂H₆ (diborane)	Dichloromethane, tetrahydrofuran	-78 to RT RT	carboxylic acid → alcohol amide → amine ketone → alcohol aldehyde → alcohol
BH₃S(CH₃)₂ (borane-dimethylsulphide complex)	Tetrahydrofuran, dimethyl sulphide	-78 to RT RT	carboxylic acid → alcohol amide → amine ketone → alcohol aldehyde → alcohol
(Sia)₂BH (disiamylborane)	Tetrahydrofuran	-78 to RT	ketone → alcohol aldehyde → alcohol lactone → lactol
AlH₃ (alane)	Ethers	-78 to RT	ester → alcohol ketone → alcohol aldehyde → alcohol alkyl halide → alkane epoxide → alcohol lactone → cyclic ether
ⁱBu₂AlH (DIBAL) (Di-isobutylaluminium hydride)	Dichloromethane, ethers, toluene	-78 -78 to RT	ester → aldehyde lactone → lactol amide → aldehyde nitrile → aldehyde ketone → alcohol aldehyde → alcohol α,β-unsaturated ketone → allylic alcohol acetylene → cis-alkene

Appendix 5 cont'd

2. Single electron transfer reducing agents

Reagent	Typical Solvents	Temperature (°C)	Functional Groups Reduced
Li/NH₃ (lithium in ammonia)	Liquid ammonia	-78 -78 to -33	α,β-unsaturated ketone → ketone ketone → alcohol aldehyde → alcohol aryl ring → dihydroaryl ring
LiC₁₀H₈ (lithium naphthalenide)	Tetrahydrofuran	-78 to RT	sulphide → alkane sulphone → alkane
NaHg (sodium amalgam)	Methanol	0-RT	ketone → alcohol aldehyde → alcohol sulphone → alkane

3. Common hydrogenation catalysts

Catalyst	Solvent	Temperature (°C)	H₂ Pressure (bar)	Functional Groups Reduced
Ni	Alcohols, toluene	5 to 100	3-10	alkene → alkane ketone → alcohol aldehyde → alcohol
Pd/C	Alcohols Alcohols, acetic acid Toluene	5 to 50 5 to 100 5 to 100	1-5 1-10 1-10	aromatic nitro → aromatic amine nitrile → amine acetylene → alkane alkyl halide → alkane

Appendix 6 : Common oxidizing reagents

Reagent	Typical Solvents	Temperature (°C)	Functional Groups Oxidized
CrO₃/H₂SO₄ (H₂CrO₄) (Jones reagent; chromic acid)	Acetone; ether	0 to RT	2° alcohol→ ketone 1° alcohol→ acid aldehyde → alcohol
CrO₃.Py₂ (Collins reagent)	Pyridine;dichloromethane	0 to RT	2° alcohol→ ketone 1° alcohol→ aldehyde
PyH⁺CrO₃Cl⁻ (PCC) (pyridinium chlorochromate)	Dichloromethane; DMF	0 to RT	2° alcohol→ ketone 1° alcohol→ aldehyde
(PyH⁺)₂Cr₂O₇ (PDC) (pyridinium dichromate)	Dichloromethane; DMF	0 to RT	2° alcohol→ ketone 1° alcohol→ aldehyde
Ag₂CO₃/Celite (Fetizon's reagent) (silver carbonate on Celite)	Hexane; benzene; chloroform	RT, Reflux	2° alcohol→ ketone diols→lactones
MnO₂ (manganese dioxide)	Hexane; benzene; dichloromethane	0, RT, Reflux	selective for allylic or benzylic alcohols→ aldehydes or ketones
KMnO₄ (potassium permanganate)	Often used in aqueous solution	0, RT, reflux	Very powerful oxidant 2° alcohol→ ketone 1° alcohol→ acid

Appendix 6 cont'd

Reagent	Solvent	Temperature	Transformation
KMnO$_4$ cont'd			alkene→ diol sulphide→ sulphone
RuO$_4$	Carbon tetrachloride/ acetonitrile/water	RT	cleaves alkenes → carboxylic acids
Al(OR)$_3$/Me$_2$CO (Oppenauer oxidation) Al(OR)$_3$/cyclohexanone	Acetone; toluene	RT Reflux	2° alcohol→ ketone
Pb(OAc)$_4$ (lead tetraacetate)	Benzene; acetic acid; acetonitrile	-78 to RT	cleaves 1,2 diols → C=O compounds ketones→ α-acetoxy-ketones many other systems also oxidized
DMSO/electrophilic reagent (El.) El. = (COCl)$_2$ (oxalyl chloride) /Et$_3$N DCC (dicyclohexylcarbodiimide) (CF$_3$CO)$_2$O (trifluoroacetic acid anhydride) Py.SO$_3$ (pyridine-sulphur trioxide)	Dichloromethane Dichloromethane Dichloromethane DMSO	-40 to RT	2° alcohol→ ketone 1° alcohol→ aldehyde
N-Chlorosuccinimide (NCS)/Me$_2$S (DMS) (N-Chlorosuccinimide/dimethyl sulphide)	Toluene	-20	2° alcohol→ ketone 1° alcohol→ aldehyde

Appendix 6 cont'd

Reagent	Typical Solvents	Temperature (°C)	Functional Groups Oxidized
OsO$_4$/N-methylmorpholine-N-oxide (NMO) (osmium tetroxide/ N-methylmorpholine-N-oxide)	Acetone/water; t-butanol	RT	alkene→ 1,2-diol
OsO$_4$/NaIO$_4$ (osmium tetroxide/sodium periodate)	Ether/water; dioxan/water	RT	cleaves alkenes→ C=O compounds
m-Chloroperoxybenzoic acid (MCPBA)	Dichloromethane	-20 to RT	alkene→ epoxide sulphide→ sulphoxide/sulphone
Ti(OPr^i)$_4$/Bu^{t}OOH/tartrate ester (Sharpless oxidation) (titanium isopropoxide/t-butyl hydroperoxide/ dialkyl tartrate)	Dichloromethane	-20	enantioselective epoxidation of allylic alcohols
VO(acac)$_2$/Bu^{t}OOH (vanadyl acetylacetonate/t-butyl hydroperoxide)	Dichloromethane	-20 to RT	allylic alcohols→ epoxides
H$_2$O$_2$/PhCH$_2$3N$^+$ OH$^-$ (hydrogen peroxide/benzyltrimethyl-ammonium hydroxide)	Alcohols	0 to RT	α,β-unsaturated C=O→ epoxide
PdCl$_2$/CuCl$_2$/O$_2$ (Wacker oxidation) (palladium chloride/cupric chloride/oxygen)	Sulpholane/water	RT to 100	terminal alkenes→ methyl ketones

Appendix 6 cont'd

Pt/O$_2$ (platinum/oxygen)	Acetone/water	RT to 100	1° alcohol→ acid diol→lactone
O$_3$ (ozone)	Dichloromethane; methanol	-78 to RT	cleaves alkenes → carbonyl compounds
TPAP/N-methylmorpholine-N-oxide (NMO) (Tetrapropylammonium perruthenate/NMO) (Pr$_4$N$^+$RuO4$^-$)	Dichloromethane	RT	1° alcohol→ aldehyde 2° alcohol→ ketone diol→lactone Cleaves 1,2-diols Sulphides→ sulphones

Index

Accidents 3-4,6-7

Acetic acid 60

Acetic anhydride 74

Acetone 56,60

Acetonitrile 56,58,60

Acetyl chloride 74

Acetylene 6,120

Acids 5

Addition funnel,
    see Dropping funnel

Adsorbent 12

Agitation 138,170,237

AIBN 75

Air sensitive reagents,
    see Reagents

Alcohols 57,58,181

Alkali metals 5-6

Alkenes 6

Alkylating agents 5-6

Alkyllithium reagents,
    see Organolithium reagents

Alumina 56

Amines 6,56,57,58,
182

Ammonia 60,180

    reactions in 247-249

Anhydrous conditions 128

Anti-bumping granules,
    see Boiling chips

Apparatus,
    see Equipment

Argon 40, 117

Aromatics, polycyclic 6

Arsenic 6

Ascarite 120

Azides 6

Balances 38

Balloons

    for hydrogenation 243

    for inert atmosphere 140-143

Barium oxide 56

Bases 5, 178

Beilstein 265,269,273

Bench space 37

Benzaldehyde 73

Benzene 6, 54, 60

Benzyl bromide 73

Bibby clips 78, 131

Birch reductions 248

Boiling chips 193,199

Boiling point 12,16,31,
250,254-
255

Boiling range 16

Borane 115

Boric anhydride 56

Boron tifluoride etherate 73,180

Bromine 6

Brookhaven Protein Databank 271

Bubblers 47,78,80,
111,131

t-Butanol 61

n-Butyllithium
    88,100,102,178

t-Butyllithium 81,91

Calcium carbide 120

Calcium chloride 56-57

Calcium hydride 56-57,72

Calcium sulfate 57

Cambridge Structural Database 270

Cannulation 78-84,134,
138-139,
143

Carbon dioxide 120

Carbon disulphide 61

Carbon monoxide 120

Carbon tetrachloride 54,61

Carcinogens 6

CAS ON-LINE 269,274

CASREACT 269,275

Celite 181,186

Characterization	13-14,17, 19,250-257	Cooling mixtures	162
		Copper(I) iodide	75
Charcoal decolorising	185	Craig tube	51,187,234
Chemical Abstracts	263-264, 269,272-275	Crotonaldehyde	72
		Crystallization	12,50,184-187,234
CAS ON-LINE	269,274	air-sensitive compounds	192-193
CASREACT	269,275	low temperature	189-192
Chemical Journals On-line	269	small scale	187-188
Chemical Literature	262	Cuprate reagents	75,98,129, 178
Chemical information	264,271		
Chemical purity	16	Current awareness	275
Chemicals	3	Cyanides	6,7
Chiral compounds	14	Cyclohexane	61,72
Chlorates	6	Danger assessment	4
Chlorinated solvents	6	Data book	8,13,19,20
Chlorine	6,113,121	Data collecting	13
Chlorobenzene	61	Data records	8,13,19-20,25
Chloroform	54,56,61, 182	Data sheet	8,13,19-23, 25,32-33
Chloroform-D, see Deuteriochloroform		Dean and Stark traps	170
Chromatography	204-226, 234-235	Decalin	62
		Decolorising charcoal	185
see also, Flash chromatography, Gc, Hplc, Mplc, Tlc		Decomposition	260
		Desiccators	58,74,130
Circular dichroism (cd)	19	Desk space	37
Clearfit apparatus	105	Detectors	153
Combustion analysis, see Microanalysis		Deuteriochloroform	61,251
		Dewar flasks	6
Computer bulletin boards	270	Dial gauges	127
Computer databases	268	Diaphagm pumps	38-39,47, 53,123-124, 207
Condensers	136,166		
air	231		
coil	136,167	Diastereoisomers	16
cold-finger	167,168	Diazo compounds	6
double-jacketed	167	Diazomethane	121
Liebig	136,167	hazards	104
Cooling baths	133,160	preparation of	103-105
dry ice-solvent	163,168	titration of	106
Ice-salt	162	Diazonium salts	6
liquid nitrogen slush baths	163	1,2-Dichloroethane	62

Dichloromethane   59,62
Diethyl ether   41,54-55,
                                        59,62-63
Diethylaluminium chloride   73
Diisopropylamine   72
1,2-Dimethoxyethane (DME)   59,63
N,N-Dimethyl formamide   57-58,63,
                                        73,182
Dimethyl sulfoxide (DMSO)   58,63,182
Dimethyl sulphate   6
Dioxan   63
Distillation   38-39,59,
                                        122,125-
                                        126,184,
                                        193-203,
                                        233
    equipment   41
    fractional   41,196-197
    fractionating columns   41
    Kugelrohr apparatus   38,202,233
    inert atmosphere   72,194-196
    normal pressure   72
    one-piece apparatus   41,72,195,
                                        200
    pig   200
    pressure-temperature nomograph
                                        199
    reduced pressure   39,73,197-
                                        200
    set-up   12
    small scale   201-203
    Spaltrohr columns   198,202
    spinning band   198,202
    solvent stills   37
    Vigreux columns   197
    medium pressure   124
Double manifold,
    see Manifold
Dreschel bottle   111
Drierite   57
Dropping funnel, jacketed   237
Dropping funnel, pressure equalizing
                                        82,137,237

Dry ice-solvent baths   163
Drying agents   31,55
Drying oven   38
Elemental analysis,
    see Microanalysis
Emergency procedures   3-7
Emulsions   181-182
Enantiomeric purity   19
Enantiomers   16
Equilibria, driving of   168-170
Equipment   36-53
    individual bench kit   36-37,44-53
    electrical   5
    general laboratory   36-44
    Quickfit   41
Esterification, procedure for   106
Ethanol   57-58,63
Ether,
    see Diethyl ether
Ethers   6,57-59
Ethyl acetate   40,64,182
Experimental data   8-9,13,33
    journal style   33
Experimental method   8,13,33
Experimental results   24
Extraction   181-182
    drying of extracts   55,182
    emulsions   181-182
Face masks   4
Filter sticks   191,194
Filtering aids   50
Filtration   50-52,229
    aids   50-52
    Craig tube   51,187,234
    filter sticks   191,194
    Hirsch funnels   50
    one-piece apparatus   50
Fire blankets   6
Fire brigade   6
Fire extinguishers   6
Fire fighting equipment   4,6
First aid equipment   4,7

Flash chromatography 32,52,205-216,234
  columns 206
  dry-column 215-6
  equipment 205
  fraction collecting 208
  glassware 52-53
  gradient elution 213
  pressure-release valve 207
  sample loading 212
  size of fractions 213
  solvent reservoir 146,206
  solvent system 208
Flash vacuum pyrolysis 247
Flash valve 53
Flex-needle 83
Fluorine 6
Fractionating column 41,197
Frit 111
Fume cupboard 37
Furniture 36
Gases 107-121
  as reagents 107,118-119
  controlled delivery of 110-111
  handling 107,110-112
  measurement of 113-117
  preparation of 119-121
  toxicity 111,112
Gas burettes 114-115,241
Gas cylinders 5,107-109
Gas scrubbers 118-119
Gas generator 117
Gas inlet T taps 49
Gc 12,19,128,134,156-159,234
  capillary columns 44,157
  compund identification 158
  column types 157

quantitative analysis 158
reaction monitoring 158
with mass specrometry 158
Generic name 32
Glassware 5,44
  assembling 130
  drying 130
  general laboratory 42
  specialized 46
Glove bag 92,97
Glove box 92
Gloves 4
Grease, for glassware 130
Grignard reactions 75
Grignard reagents 54,76,98,129,139,178
  preparation of 100
Hazardous materials 4
Hazards 3-4
  common 4-5
  explosion 6
  severe 5
Heating devices 168,237
Heavy metals 6
Hexamethylphosphoramide (HMPA) 6,57-58,64,182
Hexane 64
High pressure reactions 171
High pressure apparatus 7
High vacuum pump, see Vacuum pump
Hirsch funnels 50
Hotplate 168
Hplc 12,19,43,55,128,152-156,234
  analytical 153
  chiral 19
  columns 224
  preparative 224-226
  reaction monitoring 154

solvents for	55,226	Large scale reactions	133,138-
Hydride reducing agents	76,178		139, 236-
quantitative analysis of	115		238
Hydrocarbons	54,56-59	Lead	6
Hydrofluoric acid	6	Lecture bottles	107,110
Hydrogen bromide	121	Lewis acids	72,76,88,
Hydrogen chloride	121		180,260
Hydrogen sulphide	6	Liquid air	6
Hydrogen cyanide	6	Liquid ammonia,	
Hydrogen halides	113	see Ammonia	
Hydrogenation	113,240-	Liquid nitrogen	168
	244	Liquid nitrogen slush baths	163
catalyst	242	Liquid oxygen	6
high-pressure	244	Liquids	
low pressure	114,240	drying of	71,73
medium- pressure	244	purification of	71
via balloon	243	transfer under inert atmosphere	78
Ice-salt baths	162	Literature	
Immersion-well reactor	245	chemical	9,15
Inert atmosphere	36,48-49,	primary	263
	59,132,140-	references	12-13
	142,162	searching	26,271
Inert conditions	47,128	secondary	263
Inert gases	40,47,107,	structure of	262-263
	118,131,259	Lithium aluminium hydride	57,76,93,
Inert gas lines	122,129		96,178-179
Injuries	7	Lithium diisopropylamide (LDA)	
Internet	271		102,178
Ir	17,31-32,	Lithium amides preparation of	102
	183,253,272	Lithium amides	98,248
nujol mulls	32,253	Lithium metal	93-94,96,
Isobutyraldehyde	72		118
Isomer ratios	19	Low temperature reactions	160
Isomeric homogeneity	17	Luer adapter	77
Isomeric impurities, levels of	16	Luer to Luer stopcock	82
IUPAC Handbook	35	Magnesium sulfate	32,56-57,72
IUPAC name	32	Magnesium	57,75
Joints, Teflon sleeved	130,136	Magnetic follower,	
Koffler block	32	see Magnetic stirrer bar	
Kugelrohr apparatus	38,202,233	Magnetic fleas,	
Lab book	4,8-13,129	see Magnetic stirrer bar	
Laboratory work, records of	8	Magnetic stirrer bar	172,229
Laboratory facilities	36		

Magnetic stirrers	130,133, 171,173	Molecular formula	13,16
		Molecular sieves	56-57,171
Manifold	49	Molecular weight	12
double	39,46-48, 122,125, 130-132	Mplc	217-224, 238
		Needles	
spaghetti	48,143-144	double ended	78
Manometer	126	disposable	87
Mass spectrometry	16-17,254	fixed	85
chemical ionisation (CI)	16-17,32	flat ended	87
electron impact (EI)	16-17,32	removable	85
electrospray	17	Luer fitting	87
fast atom bombardment (FAB)	17	screw-on	86
fragment ions	18	Needle valve	108,110
molecular ion	16-17	Nitrates	6
pseudo molecular ion	16	Nitro compounds	6
spectra	17,32,250, 254,272	Nitrogen	40,118
		Nitromethane	64
McLeod gauge	127	Nmr	55,61,159, 183,251-253,272
Mechanical shakers	175		
Mechanical stirrers	173		
Melting point	12,16,31, 254-255	assignment	17
		2-dimensional	14,18
determination of	32	^{13}C	18,31,251
Melting point apparatus	32,254	chemical shift	17-18,32
Mercury bubbler	114,117	chiral shift reagent	19
Mercury	6	choice of solvent	251-252
Metal acetylides	6	coupling constants	14,17
Metal dispersions	93-95	decoupling	14
Metal fluorides	6	^{1}H	17,31
Metal hydrides	54,76,93-94,97,129	multiplicity	17
		nOe experiments	14,18
Methanol	57-58,64	spectrum	14
Methyl iodide	6,72	techniques	18
Methyl lithium	178	NOe experiments, see Nmr	
Methylene chloride, see Dichloromethane		Nomograph	199
		Notebook format	10,12
Microanalysis	15-16,55, 184,250, 256-257	Optical rotation	18,255-256
		Optical rotatory dispersion (ord)	19
Mininert bottles	91		
Mininert valves	76,91	Organoaluminium compounds	76,129, 178

Organoboranes	76,129		199
Organolithium reagents	54,76,88,	Product, isolation of	181
	98,129,139	Purification	14,183-226,
preparation of	100		233-235,
titration of	102		237-238
Organometallic reagents	178	Purity	13,15,17,19
preparation of	98	Pyridine	182
titration of	98	Pyridines	56-58,64
Osmium tetroxide	6	Quickfit equipment	41
Oxalic acid	6	Raman	272
Oxalyl chloride	6	Raney nickel	243
Oxidizing agents	5	REACCS	270,275
Ozone	6,246	Reactions	
Ozonides	246	above room temperature	135,163-
Ozonolysis	246		167,231-
Papers	9,13		232
Parafilm	142-143	anhydrous	36,47
Pentane	64	agitation	170-176,
Perchlorates	6		237
Perchloric acid	6	heating	237
Perkin triangle	201	inert atmosphere	129-143
Peroxides	6,54,56,62	in nmr tubes	232-233
Personal items	45	large scale	137,236-
Petrol,			239
see Petroleum ether		low temperature	160-163,
Petroleum ether	40,64		228-231
Phosgene	6	monitoring	12,134,144-
Phosphorous pentoxide	56,58,130		159
Photochemical reactor	245	quenching	177-178
Photolysis	244-246	sealed tube	164,231
low pressure lamps	245	small scale	138,227-
medium pressure lamps	245		235
Pirani gauges	127	work-up	12,135,177-
Potassium hydroxide	58		183
Potassium hydride	93-94,96,98	yield	13
Potassium	93	Reaction tube	165
Pressure	12	Reagents	69
apparatus	5	addition of	134,137-
measurement	126		138
units of	127	air sensitive	69,76,128-
Pressure gauge	110		129
Pressure regulator	108-109	drying	69,71-75
Pressure-temperature nomograph		handling	70,76-98

measuring	76-98
moisture-sensitive	69
purification	69,71-75
storing under inert atmosphere	
	76-78
toxic	70
unstable	70
Recrystallization,	
see Crystallization	
Refrigeration unit	164
Refrigerator	38
Regulations	4
Reports	9,13,27,33
Reservoirs	52
Retention times	19
Rodaviss joints	52,207
Rotaflow tap	79
Rotary evaporators	32,38,124
Rotary evaporation	39,122
Rotary oil pumps	124
Safe working practice	4
Safety	3-7,128
Safety audit	4
Safety equipment	4
Safety legislation	4
Safety shields	4
Safety spectacles	3-4
Sand buckets	6
Schlenk tube	81
Science Citation Index	266,269
Sealed tube	164,231
Seed crystals	185-186
Selenium	6
Septa	78
Silica dust	208
Silica Recycling	214
Silica gel self-indicating	130
Silica powder	205
Silica tlc plates	146
Sintered funnel	50-51
Sodium sulfate	59,179
Sodium amide	248

Sodium hydride	93-94,96,
	259
Sodium metal	59,93
Sodium-benzophenone ketyl	59
Sodium-D line	18,256
Sodium-potassium alloy	59
Solids	
as reactants	133
addition of a to a reaction	140
handling under inert atmosphere	
	92
weighing reactive powders	96
Solvents	5
drying	55-65
grades	54-55
polar aprotic	182
purification	54-68
stills	40,65-67
storage under inert atmosphere	
	76,78
transfer of	84
traps	47,124
Sonication, see Ultrasound	
Soxhlet extractor	171
Spaltrohr columns	198,202
Specific rotation	18,256
Spectra	13
storage	37
assignments	13
Spectroscopic data	9,13,15,19
Spectroscopic solvents	55
Spinning band columns	198,202
Stereochemisty	14
Stirrer bar,	
see Magnetic stirrer bar	
Stopper, Teflon	72,76
Structure determination	14-15,18
Sublimation	184,203-
	204
Suck-back, prevention	123
Synthetic methods, information sources	
	274

Syringes	44,84-92, 129	Titanium tetrachloride	72,88, 129,180
all-glass Luer fitting	85-86	Tlc	10,12,32, 43,128,134, 143,145- 152,177, 181,183, 208,213,260
choice of	84		
cleaning and care	87-88		
disposable	86		
fittings	86		
gas-tight	84-86		
liquid-tight	84	analytical	205
Luer-lock	87	compound identification	145
micro	84-85,130	indication of purity	146
needles, see Needles		multiple elutions	151
use of	84-92,129	plates	12,31,146
Syringe pump	138	preparative	217
Teflon sleeves	99,130	reaction monitoring	145-152
Teflon taps	49,130	Rf values	151
Teflon tape	143	running	146,151
Test tube	230	solvent system	12,150
Tetrahydrofuran (THF)	41,59,65, 181,259	spot visulization	43,148
		spotters	147
Tetralin	65	stains	12,43,148, 149
Thallium	6		
Thermometer		tank	148
digital	160	two-dimensional plates	151
low temperature	160	Toluene	65,171
Thesis	9,13,15,24- 35	p-Toluenesulphonyl chloride	75
		Toxic compounds	6
discussion	24	Triethyl aluminium	73
experimental section	20,24-25, 31-32	Triethylamine	72,182
		Trouble Shooting	258-261
introduction	24	Tubing	47,112
planning	24-26	Ultrasonic bath	176
practical procedure	12	Ultrasonic probe	177
preparation	24	Ultrasound	176
proton nmr data	17	Ultraviolet radiation	244
reference list	27	Ultraviolet spectroscopy	19,55,159, 254,272
sections	24		
writing	20,26-27, 29-30	Ultraviolet lamp	32,43,148
		Uniform resource locator (URL)	
Thin layer chromatography, see Tlc			271
		Vacuum apparatus	5-6
Tin tetrachloride	73	Vacuum desiccator	74,130
		Vacuum oven	38,74

Vacuum pumps	38,122-127, 131
diaphragm pump	38-39,47, 53,123-124, 207
house systems	38,122,131
oil pumps	39
rotary pump	47,124-126
two stage rotary pump	39
water aspirator	38,123
vapour diffusion pumps	126
Vigreux column	196
Waste chemicals	4
Water aspirators	38,123
Water baths	168
Water, continuous removal of	170
Woods metal	168
Word-processing package	19
World Wide Web (WWW)	270
X-ray analysis	14
X-ray structures	270
Xylene	65
Zinc	75